MW00824639

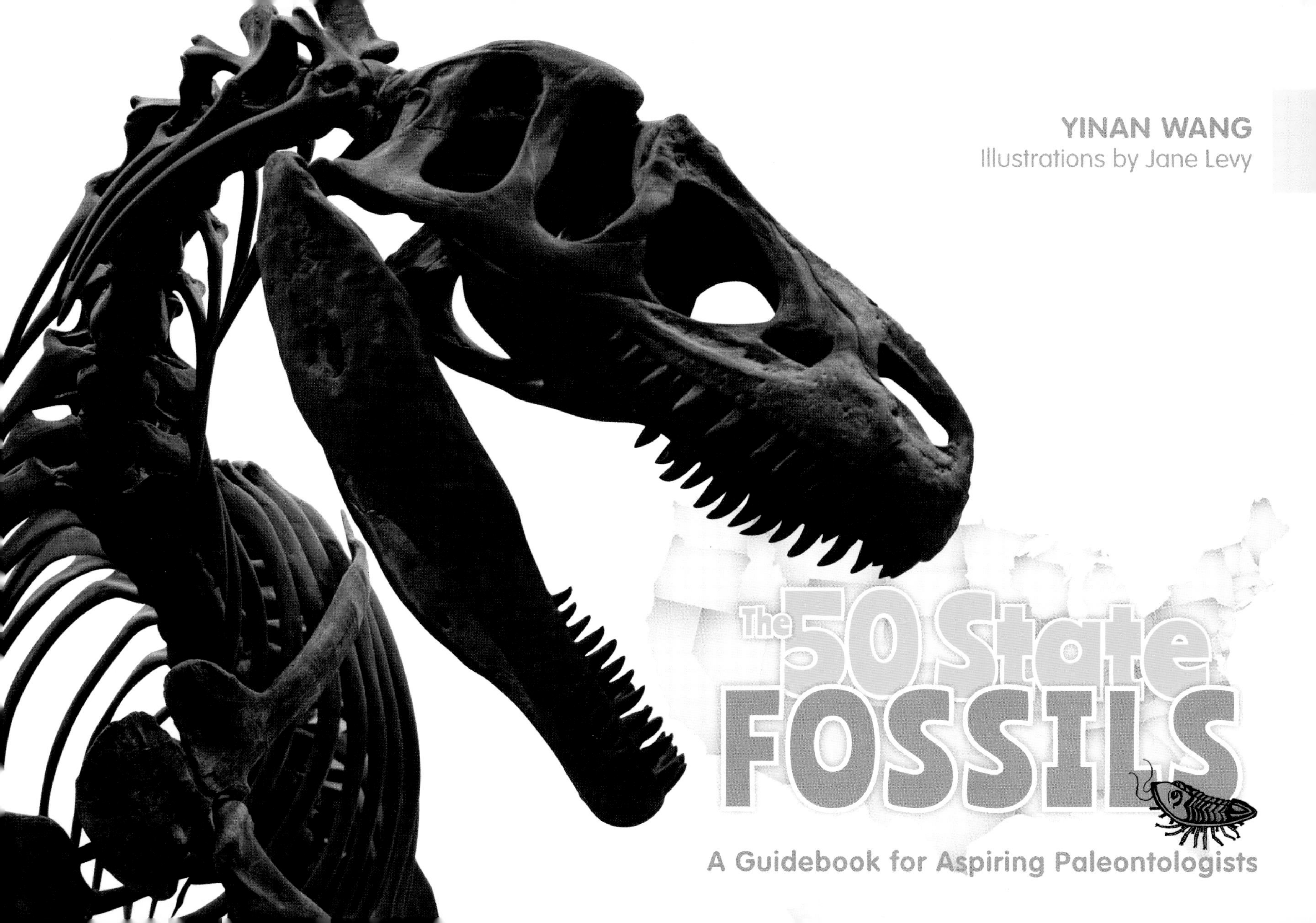
YINAN WANG
Illustrations by Jane Levy
The 50 State
FOSSILS
A Guidebook for Aspiring Paleontologists

To my parents, who took me rock collecting when I was a kid and encouraged my interest in nature and paleontology. –Yinan Wang

To my lovely husband Mike, who supports all my crazy ideas. –Jane Claire Levy

Copyright © 2018 by Yinan Wang
and Jane Levy

Library of Congress Control Number: 2018937047

All rights reserved. No part of this work may be reproduced or used in any form or by any means—graphic, electronic, or mechanical, including photocopying or information storage and retrieval systems—without written permission from the publisher.

The scanning, uploading, and distribution of this book or any part thereof via the Internet or any other means without the permission of the publisher is illegal and punishable by law. Please purchase only authorized editions and do not participate in or encourage the electronic piracy of copyrighted materials.

"Schiffer," "Schiffer Publishing, Ltd.," and the pen and inkwell logo are registered trademarks of Schiffer Publishing, Ltd.

Art Direction & Layout by
Danielle D. Farmer
Cover design by Danielle D. Farmer

Photograph of New Mexico state fossil Coelophysis ©Bailey Library and Archives, Denver Museum of Nature & Science

Type set in Ranchers/Vag Rounded

ISBN: 978-0-7643-5557-8
Printed in China

Published by Schiffer Publishing, Ltd.
4880 Lower Valley Road
Atglen, PA 19310
Phone: (610) 593-1777;
Fax: (610) 593-2002
E-mail: Info@schifferbooks.com
Web: www.schifferbooks.com

For our complete selection of fine books on this and related subjects, please visit our website at www.schifferbooks.com. You may also write for a free catalog.

Schiffer Publishing's titles are available at special discounts for bulk purchases for sales promotions or premiums. Special editions, including personalized covers, corporate imprints, and excerpts, can be created in large quantities for special needs. For more information, contact the publisher.

We are always looking for people to write books on new and related subjects. If you have an idea for a book, please contact us at proposals@schifferbooks.com.

CONTENTS

INTRODUCTION

What is it about fossils that make them special? Fossils are like time machines; they give us glimpses of the past, filled with fantastic creatures both strange and familiar. They are special objects that inspire us to think about life and all its forms. They make us feel a combination of wonder, delight, and awe. Fossils can be found in many places and yet are hard to find. Every state has fossils, and this book is about the celebrated natural treasures that have been designated as state fossils.

WHAT ARE FOSSILS?

Fossils are naturally preserved remains and traces of prehistoric life. Remains are things like shells, dinosaur bones, and leaves. Traces are preserved indications of activities such as dinosaur footprints and insect bite marks on leaves. Remains and traces must be ancient, at least 11,700 years old, to count as fossils. Objects that were intentionally made by people, such as arrowheads and pottery, are artifacts and not fossils. The study of fossils is part of a science called paleontology. A scientist who works in paleontology is called a paleontologist.

HOW DO FOSSILS FORM?

BURIAL IS USUALLY THE FIRST STEP TO BECOMING A FOSSIL.

Almost all fossils had to be buried in one way or another. A dinosaur could die in a swamp and sink into the mud. A bug could be trapped in sediment that falls on it. A forest might be blanketed in ash from a volcanic eruption. A dinosaur footprint on a river bank could harden and be covered over. These are all ways that organisms and their traces get buried.

Occasionally there are other ways an organism starts on the path of becoming a fossil. One is to get trapped in tree resin, which will become amber. Another is to get stuck in tar. Dry and frozen environments can also preserve remains of life for tens of thousands of years.

AFTER BURIAL, MANY PROCESSES CAN TURN REMAINS INTO FOSSILS; HERE ARE THE MORE COMMON WAYS THIS HAPPENS:

Cast and mold. An organism can dissolve or disintegrate after getting buried, leaving behind a hole that is called a mold. If the mold is then filled with other minerals or mud that hardens

Fossil fern cast (bottom) and mold (top)

over time, a cast of the original fossil is formed. Sometimes hollows (such as the inside of a skull) within a buried fossil are filled and this forms what is called an endocast.

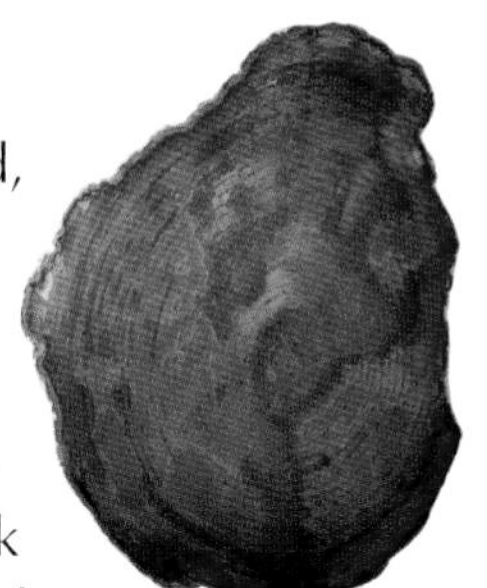
Petrified wood preserved by permineralization

Permineralization: When an organism gets buried, sometimes mineral-rich water passes through tiny spaces and cells within the organism, leaving behind minerals. These minerals help preserve a fossil by making it stronger than the original remains. When the cells of a tree trunk are filled by minerals, the result is petrified wood. Colorful dinosaur bones are often another result of permineralization.

Fossil trilobite with pyrite mineral replacement

Mineral replacement: Sometimes parts of a buried organism get replaced by minerals. For example, a shell can be replaced by the mineral pyrite (fool's gold) after burial, resulting in a shiny metallic fossil. This process is different from permineralization because the remains of an organism are getting replaced by minerals, rather than minerals filling in spaces within the remains.

Fossil fern preserved by carbonization

Carbonization: All forms of life contain the element carbon, and sometimes the carbon is all that is left of an organism. Dark carbon impressions of leaves on a rock are an example.

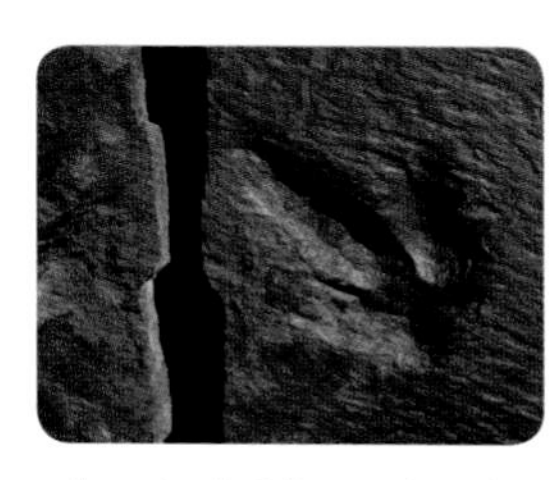
Trace fossil of dinosaur footprint

Trace fossils: Evidence of activities of an organism that become preserved are called trace fossils. Dinosaur footprints are a common example. Fossil coprolites, which are dinosaur dung, are another example. Damage to a fossil leaf from an insect feeding on it counts as a trace fossil too.

Original remains: The original remains can be preserved in special situations. Insects trapped in amber can sometimes still have original pieces of their exoskeleton even millions of years later. Fossil shells can have original mother-of-pearl. A mammoth can be frozen in permafrost with actual hair and meat preserved. Animals trapped in tar can have original bone material. Mummified animal remains have been found in caves. All these examples, if they are more than 11,700 years old, count as fossils.

Fossil insects in amber may have original remains

WHY STATE FOSSILS?

States designate official symbols to recognize and celebrate important cultural and natural objects of that state. The most common state symbols include state flags and state seals, but there are often also state birds, state flowers, state trees, state beverages, and a variety of other symbols. Designating a state fossil can bring tourism and attention to the selected fossil while encouraging an interest in the field of paleontology. While fossils occur in every state, not every state has an official state fossil. At the time of this writing there are forty official state fossils. Of the ten remaining states, some have other symbols such as state stones or state dinosaurs that also qualify as fossils, and they are included in this book. Currently several states do not have state fossils, and citizens of those states can work to get them designated.

HOW A STATE FOSSIL GETS DESIGNATED:

1. Someone notices his or her state does not have a state fossil.
2. Either individually or with a group, the citizens of that state research local fossils and pick one to propose as the state fossil. This is often done by classrooms of students but can be the effort of a single person.
3. The individual or group contacts elected state representatives and works with them to write a bill designating the state fossil.
4. The bill goes through the state legislative process and is voted on, usually by both the House and Senate of that state.
5. If the bill makes it through the legislature and is signed by the governor, the state fossil is then designated by law.

State Fossil
State Dinosaur (No State Fossil)
State Stone (No State Fossil)
No State Fossil
State Capital

FOSSIL HUNTING

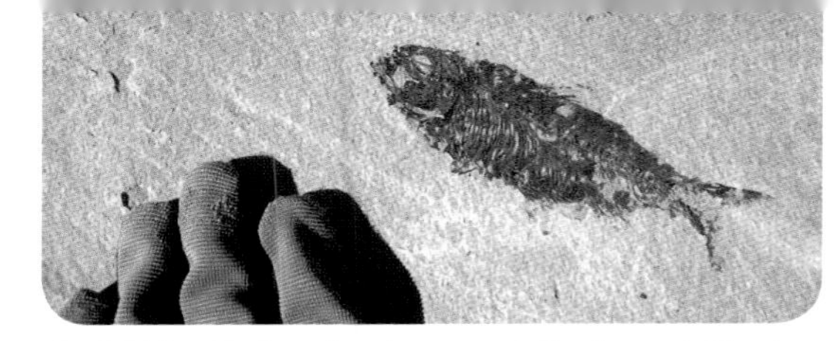

Fossil fish found by the author at a commercial fossil quarry in Wyoming

Fossil hunting is a classic hobby that involves looking for and collecting fossils. The best way to get introduced to fossil hunting is to join a local rock collecting club and attend their field trips. Fossils are found in every state, but not everywhere in a state. Clubs and other groups can help provide access to sites, guidelines for best practices for collecting, and tips on staying safe. An additional option is that many museums and universities have field trips where fossil hunters can assist paleontologists on their digs.

Clubs and other groups can also help keep fossil hunters updated about local, state, and federal laws on fossil collecting. There are rules regarding where and what a person can collect. Fossil collecting is illegal in national parks and national monuments. Many commercial fossil quarries will allow visitors to collect fossils for a fee. Fossil hunters must always ask permission before collecting on private land. Laws vary in other places and are constantly changing, so always check with the authorities of a place before planning a trip. Occasionally something rare or new to science is found and should be donated to a museum; many great scientific finds have been made by amateur fossil hunters! A person does not have to be a paleontologist to hunt for fossils, but fossil collecting is a good way to get started in becoming a future paleontologist.

HOW DO WE KNOW HOW OLD FOSSILS ARE?

Fossils and the rocks they're in sometimes contain certain elements that are unstable. Scientists know how fast these elements decay and can use that information to calculate when a fossil or rock was deposited. This is called absolute dating.

After determining the age of certain rocks, fossils in layers above and below them can be dated. Typically the layers under the dated layer are older. Layers above are younger. This is called relative dating.

Using these two techniques, scientists can figure out approximately how old a fossil is.

GEOLOGIC TIME SCALE

The Earth is approximately 4.54 billion years old. The geologic time scale was created as a way to illustrate and separate sections of that time based on events in Earth´s history. The geologic time scale is divided into the following:

EON: The longest section of time. Includes:

Phanerozoic: most recent 541 million years, when most life exists.
Proterozoic: from 2.5 billion years ago to 541 million years ago. The start of complex single-celled and multicellular life.
Archean: from 4 billion to 2.5 billion years ago. Eon when life started.
Hadean: from 4.54 billion to 4 billion years ago. When Earth formed.

ERA: Section of time within an Eon, separated to show great periods of life ended by mass extinctions.

This book focuses on the eras of the Phanerozoic Eon:
Cenozoic: 66 million years ago to today. After extinction of dinosaurs, known for the variety of mammals.
Mesozoic: 252 to 66 million years ago. Age of the dinosaurs until the Cretaceous (K)-Tertiary (T) mass extinction (also known as the K-T Mass Extinction).
Paleozoic: 541 to 252 million years ago. Beginning of complicated life, up to Permian extinction.

PERIOD: Section of time within an Era, usually ended by a major extinction that shakes up the life of the time.

Quaternary: 2.58 million years ago to now. Includes the Ice Age.
Neogene: 23 to 2.58 million years ago. Lots of mammals.
Paleogene: 66 to 23 million years ago. Evolution of a variety of mammals.
Cretaceous: 145 to 66 million years ago. Dinosaurs including Tyrannosaurus rex and Triceratops.
Jurassic: 201 to 145 million years ago. Dinosaurs including Allosaurus and Stegosaurus.
Triassic: 252 to 201 million years ago. Earliest dinosaurs appear, such as Coelophysis.
Permian: 298 to 252 million years ago. Large insects, amphibians, last of trilobites.
Carboniferous: 358 to 298 million years ago. Lots of plants started forming. The Carboniferous is also usually divided into two subperiods: Pennsylvanian, 323 to 298 million years ago; and Mississippian: 358 to 323 million years ago.
Devonian: 419 to 358 million years ago. Lots of early jawed fish and trilobites.
Silurian: 443 to 419 million years ago. Variety of sea life including sea scorpions, trilobites, and crinoids.
Ordovician: 485 to 443 million years ago. Variety of invertebrate life.
Cambrian: 541 to 485 million years ago. First appearance of many classes of life.

EPOCH: section of time within a Period. While many Periods are familiar, some Periods are divided into Epochs that are more familiar, and these are the more common ones that will be used in the book:

Quaternary period:
Holocene: 11,700 years to now.
Pleistocene: 2.58 million years ago to 11,700 years ago.

Neogene period:
Pliocene: 5.33 to 2.58 million years ago.
Miocene: 23 to 5.33 million years ago.

Paleogene period:
Oligocene: 33.9 to 23 million years ago.
Eocene: 56 to 33.9 million years ago.
Paleocene: 66 to 56 million years ago.

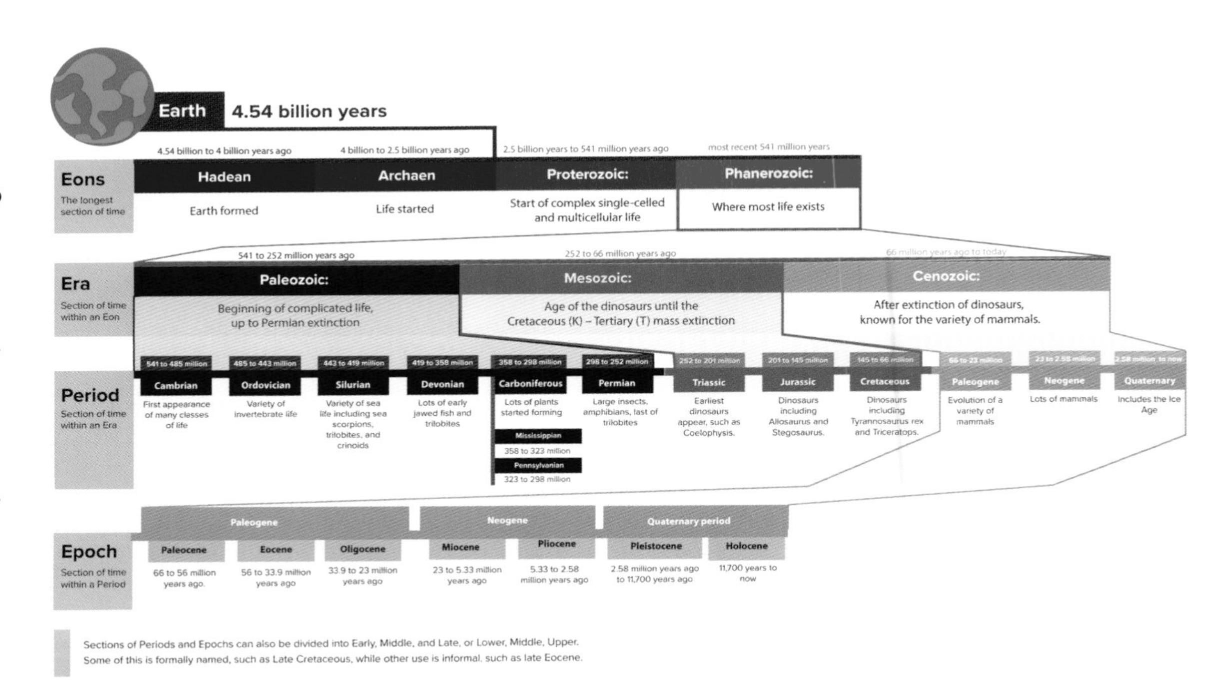

When talking about time, periods and epochs can also be divided more generally into Early, Middle, and Late, or Lower, Middle, and Upper. Some of these are formally named, such as Late Cretaceous, while other use is informal, such as late Eocene.

TAXONOMIC RANK

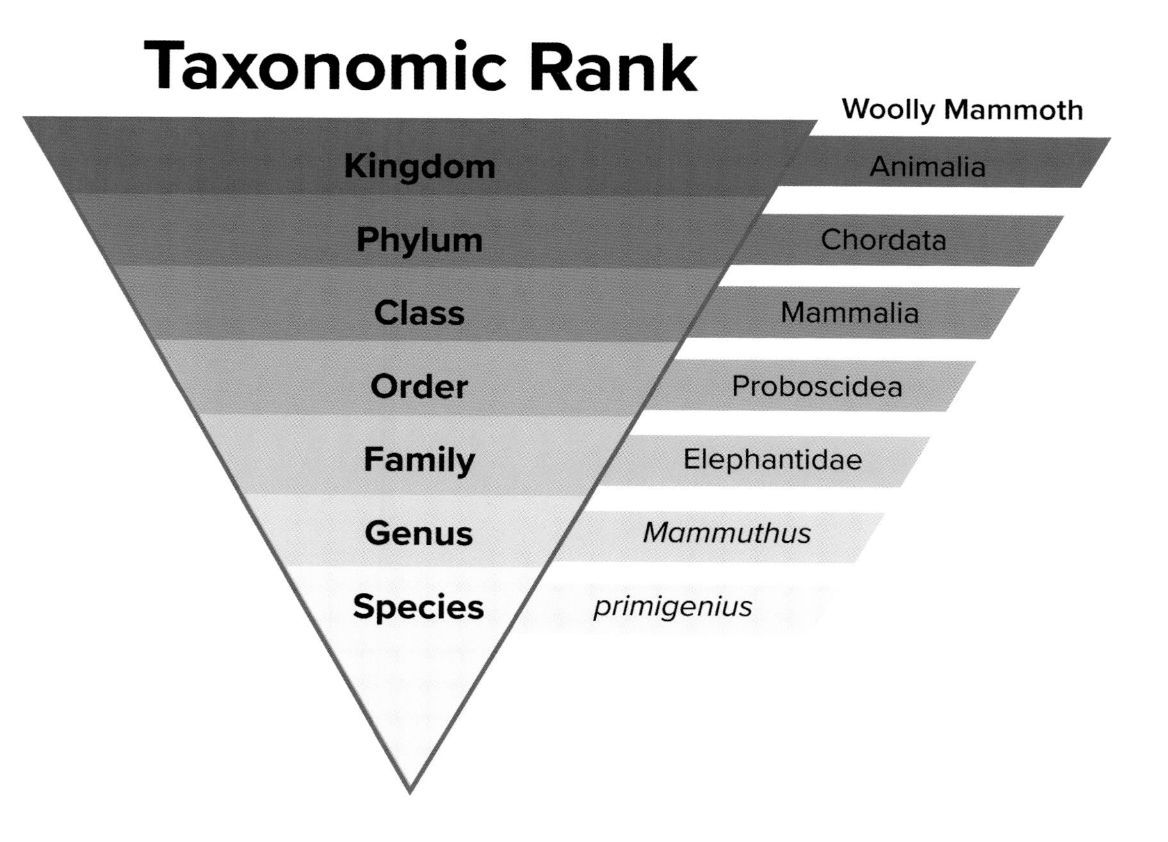

Science provides order to organisms by using taxonomic rank. The state fossils range from phylum all the way down to subspecies, and the following will help explain the position of each rank:

Kingdom: Largest group.

Phylum: Group of organisms with similar traits within a Kingdom.

Class: Group of organisms with similar traits within a Phylum.

Order: Group of organisms with similar traits within a Class.

Family: Group of organisms with similar traits within an Order.

Genus: Group of organisms with similar traits within a Family.

Species: Group of same organisms with similar traits within a Genus.

Alabama

BASILOSAURUS

Basilosaurus cetoides
(pronounced Bah-sil-oh-saw-rus set-oh-ee-dees)

Late Eocene epoch (40 to 34 million years ago)
The Basilosaurus was designated the state fossil in 1984.

Basilosaurus skeleton at the Alabama Museum of Natural History

Swimming the oceans of the late Eocene epoch, the Basilosaurus was an early toothed whale that grew up to 59 feet long. Fossils were first discovered in Alabama in 1833, and a complete skeleton was unearthed in southwest Alabama in 1834. The massive bones were first thought to be those of a giant lizard and mistakenly it was named Basilosaurus, which meant "king of lizards." Famous paleontologist Richard Owen wanted to rename it Zeuglodon, meaning "yoked tooth," but the name Basilosaurus is still used today. When it was designated the state fossil in 1984, it was also written into law that removal of *Basilosaurus cetoides* fossils from Alabama is illegal without written approval by the governor.

Alaska

Late Pleistocene to Early Holocene epoch (100,000 to about 10,000 years ago)
The woolly mammoth was designated the state fossil in 1986.

WOOLLY MAMMOTH

Mammuthus primigenius
(pronounced Ma-muh-th-us prim-i-jen-ee-us)

Standing up to 11 feet tall and weighing up to 6 tons, the woolly mammoth roamed much of North America during the Ice Age. The woolly mammoth looked like a large elephant covered in fur, with long curved tusks. They are believed to have descended from the Steppe Mammoth in East Asia 400,000 years ago and crossed a land bridge from Siberia to Alaska during the Ice Age 100,000 years ago. Mammoths went extinct about 10,000 years ago, although some survived for a few thousand more years on islands. Many woolly mammoth fossils are found trapped in Alaska's permafrost and are often found by gold prospectors who melt the permafrost to find gold. Mammoth teeth and tusks are often the best preserved fossils due to their hardness.

Fossil woolly mammoth skeleton

Arizona

Phoenix

State Capital

State fossils may be found in this area

PETRIFIED WOOD

Araucarioxylon arizonicum
(pronounced Ah-raw-car-i-ox-zy-lon ari-zon-i-com)

Late Triassic period (227 to 205 million years ago)
Petrified wood was designated the state fossil in 1988.

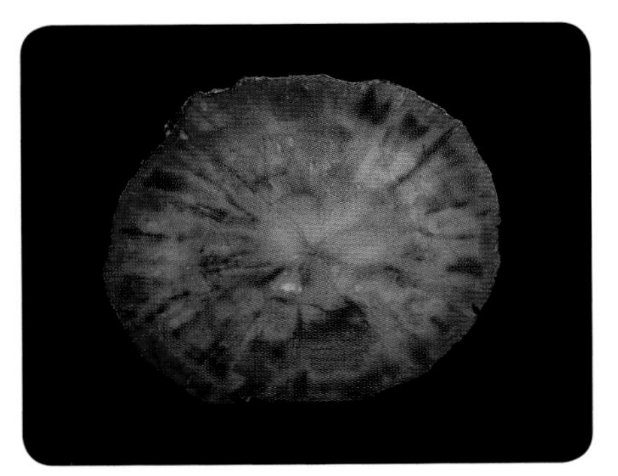

Slice of petrified wood from Arizona

During the late Triassic, Northern Arizona was covered by forests of conifer trees. These massive pine trees called *Araucarioxylon arizonicum* grew up to 200 feet tall and had trunks up to 5 feet thick. Occasionally fallen trees would be buried in sediment and volcanic ash. Mineral-rich water would get absorbed by the tree's wood and deposit hard minerals, a process called permineralization, resulting in petrified wood. Most of Arizona's petrified wood is colorful because it absorbed a lot of silica and became varieties of colorful quartz. Some of the best petrified wood can be seen in Petrified Forest National Park, which was established as a national monument in 1906 to protect thousands of acres containing petrified wood. It became a national park in 1962.

Arkansas

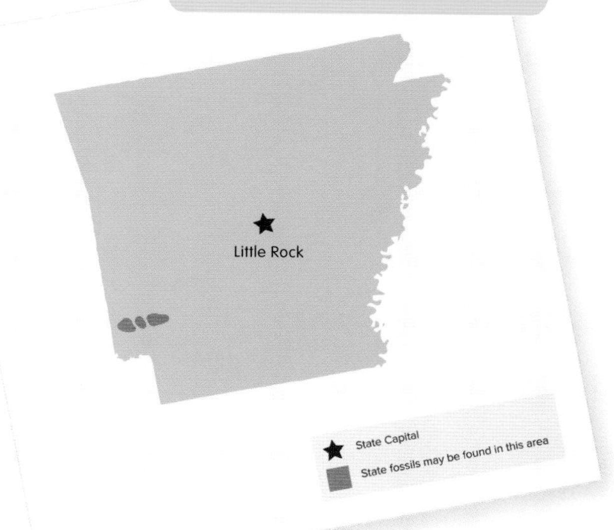

Early Cretaceous period (132 to 105 million years ago)
The Arkansaurus was designated the state dinosaur in 2017.

ARKANSAURUS

Arkansaurus fridayi
(pronounced Ar-kan-saw-rus frie-day-eye)

Arkansas does not have a state fossil but does have a state dinosaur. In 1972, a partial dinosaur foot was found in a gravel pit near Lockesburg, Arkansas. The identity of the dinosaur was unknown at the time and it was informally named *Arkansaurus fridayi to* honor the person who found it, Joe B. Friday. It wasn't until 2018 that the name *Arkansaurus fridayi* became official after a full scientific study was published on the foot. The Arkansaurus was a theropod dinosaur from the Ornithomimosauria (pronounced Or-nif-o-mime-o-sore-ee-a) clade, which was a group of slender fast-moving dinosaurs that loosely resembled ostriches. These are the only known dinosaur remains from Arkansas so far; however, there are dinosaur tracks in Southwest Arkansas.

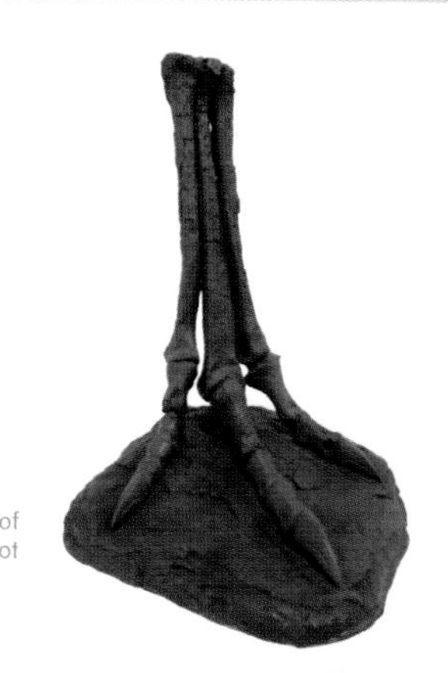

Model of
Arkansaurus foot

California

Note:
The ***Augustynolophus morrisi*** was designated the official state dinosaur in 2017. It was a hadrosaur (a duck-billed dinosaur) and has only been found in California.

SABER-TOOTHED TIGER

Smilodon californicus
(pronounced Smie-loh-don cal-if-or-nik-us)

Late Pleistocene epoch (1.6 million years ago to 11,700 years ago)
The Smilodon was designated the state fossil in 1973.

Smilodon skull featuring long canine teeth

Saber-toothed cats are ferocious felines that existed from 42 million years ago to up to 11,700 years ago. They had distinctive saber-shaped upper canine teeth that extended from the mouth even when it was closed. The most famous of the Saber-toothed tigers was the *Smilodon californicus* (a synonym of the more familiar name *Smilodon fatalis*), which roamed North America during the Ice Age up to 11,700 years ago. The Smilodon grew up to 600 pounds and had saber-shaped canines up to 7 inches long. More than 100,000 bones from *Smilodon californicus* have been found in the La Brea Tar Pits in Los Angeles, California. The tar pits trapped many animals, which then helped attract hungry Smilodon that also ended up getting stuck and preserved in the tar.

Colorado

Late Jurassic period (157 to 150 million years ago)
The Stegosaurus was designated the state fossil in 1982.

STEGOSAURUS

Stegosaurus genus
(pronounced Steg-ah-saw-rus)

First discovered in Colorado in 1877, the Stegosaurus was a four-legged plant-eating dinosaur that roamed North America during the Late Jurassic. The most recognizable features of the Stegosaurus are the upright bone plates along its back and several large spikes on its tail. The purpose of the plates has been much debated; they might have been used for display or to help control the dinosaur's body temperature, or maybe both. The large tail spikes, which measured up to 3 feet long, were used for defense against other dinosaurs. The Stegosaurus could grow up to 29 feet long and weigh up to 3 tons. In the 1880s, a paleontologist made a cast of a well-preserved Stegosaurus brain case and found that the brain was little bigger than a walnut.

Stegosaurus skeleton with iconic plates and tail spikes

Connecticut

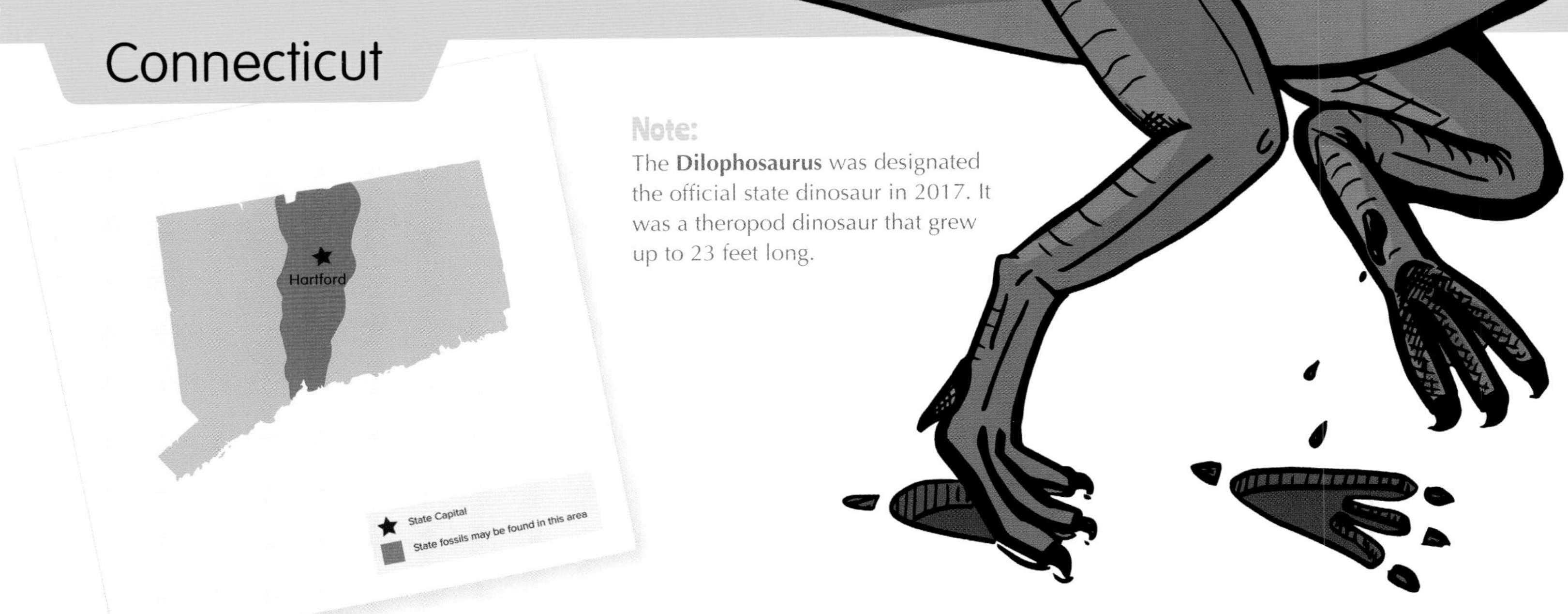

Note:
The **Dilophosaurus** was designated the official state dinosaur in 2017. It was a theropod dinosaur that grew up to 23 feet long.

DINOSAUR TRACKS

Eubrontes ichnogenus
(pronounced Yoo-bron-tees)

Early Jurassic period (about 200 million years ago)
The Eubrontes was designated the state fossil in 1991.

Example of Eubrontes at Dinosaur State Park

When fossil tracks were first discovered in the Connecticut River Valley of Massachusetts in the 1800s, they were believed to belong to a giant bird. They were given the name Eubrontes, which is an ichnotaxon, a term for a group of the same trace fossils of an organism. By the 1950s, it was agreed that the tracks were made by a large carnivorous theropod dinosaur, but the dinosaur has not been identified. The tracks measure up to 19 inches long with sharp claw marks. A dinosaur like the Dilophosaurus (pronounced Dil-of-oh-saw-rus) might have made them; however, no remains have ever been found with the tracks. In 1966, construction in the town of Rocky Hill, Connecticut, uncovered a significant number of Eubrontes dinosaur tracks. This site became Dinosaur State Park in 1968, and features more than 2,000 dinosaur tracks.

Delaware

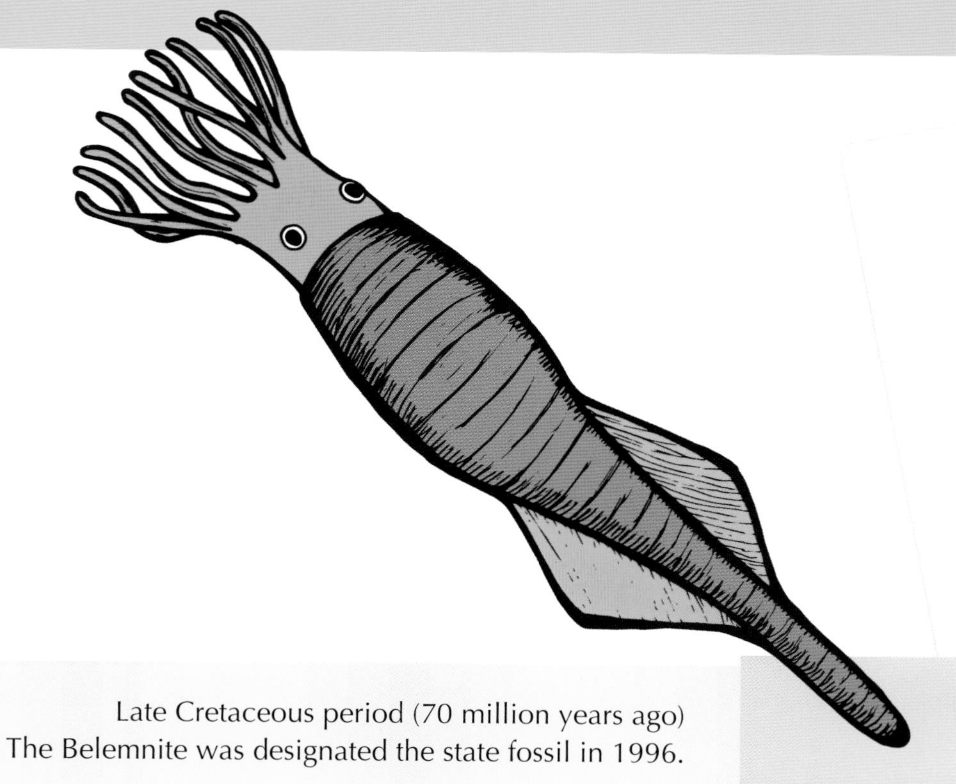

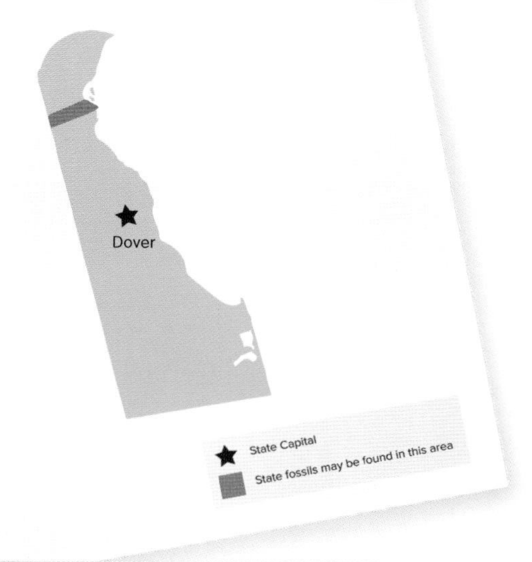

Late Cretaceous period (70 million years ago)
The Belemnite was designated the state fossil in 1996.

BELEMNITE

Belemnitida order
(pronounced Bell-em-ni-ti-dah)

Belemnites were cephalopods, a class of animals that includes octopuses and squids, and they lived in the oceans throughout the world from the Jurassic to the end of the Cretaceous. Belemnites looked similar to squids and had 10 long arms that were covered in hooks to grab prey. They had an internal hard structure that served as a skeleton. The rostrum was a part of this internal structure, and because it was composed of hard calcite, it is the most common fossil remains of Belemnites. The construction of the Chesapeake and Delaware Canal excavated a lot of sediments from the Late Cretaceous age Mount Laurel formation, where numerous fossils of the species *Belemnitella americana* are found today.

Fossil Belemnite rostrum

Florida

AGATIZED CORAL

Anthozoa class
(pronounced An-tho-zoh-ah)

Oligocene to Miocene epoch (33 to 20 million years old)
Agatized coral was designated the state stone in 1979.

Fossil coral from Florida with nice colors

Florida does not have a state fossil but does designate fossil agatized coral as the state stone. Corals are colonies of marine animals of the class Anthozoa. They colonize surfaces as individual polyps, which are tiny invertebrates. The coral polyps make an exoskeleton of calcium carbonate, resulting in a hard structure that accumulates into coral reefs. When coral is buried, it can be fossilized via mineral replacement as the calcium carbonate is replaced by silica to form the mineral chalcedony. Agate is a type of chalcedony, and agatized corals of various colors are found in many places in Florida, generally being 33 to 20 million years old.

Georgia

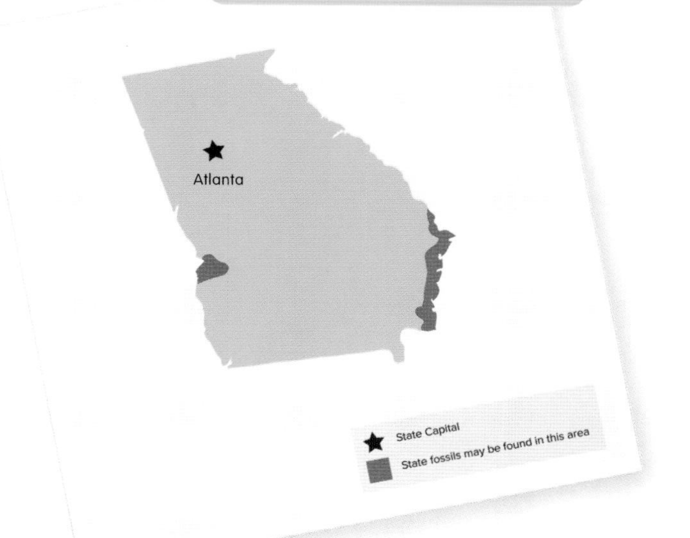

Cretaceous period to Pleistocene epoch (87 million to 11,700 years ago)
The shark tooth was designated the state fossil in 1976.

SHARK TEETH

Elasmobranchii subclass
(pronounced Ee-las-mo-br-on-kee-i)

Sharks are predatory fish that have been swimming the oceans for more than 420 million years. Sharks constantly replace their teeth, and some can go through more than 30,000 in a lifetime. Fossilized shark teeth are strong and can survive being washed around in water better than other fossils. Because of how numerous and strong they are, shark teeth are very common fossils and can be found on beaches and in rivers. They range in size from a grain of sand up to 7 inches. Georgia counts all types of shark teeth as the state fossil.

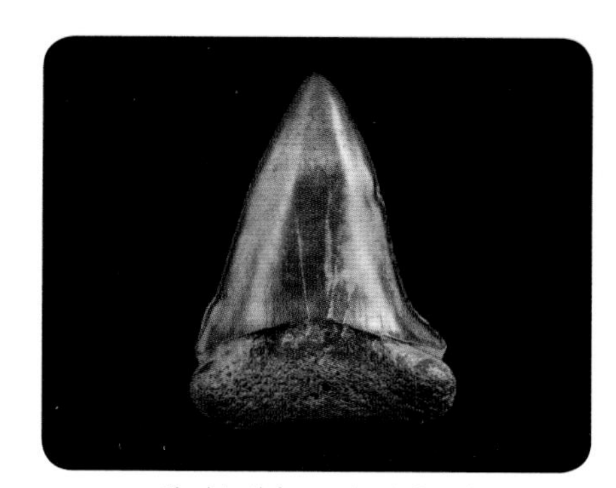

Shark tooth from a river in Georgia

Hawaii

HAWAII

does not have a state fossil

Moa-nalo skull found in a lava tube on Maui

Recommendation for state fossil:

Hawaii is geologically new compared to other states, with most of the larger islands having been formed within the past 5 million years. There are some Pleistocene reef related fossils in various places, including Oahu. Hawaii also has a variety of extinct birds, and their rare fossilized or near-modern bones are occasionally found in sand dunes and caves.

One of the more interesting extinct birds is the Moa-nalo, a goose-like flightless bird that went extinct after humans arrived on the islands more than a thousand years ago. The Moa-nalo evolved from small ducks that arrived on the islands several million years ago. Over time they evolved to be bigger, lost the ability to fly, and became one of the major herbivores of the islands.

Idaho

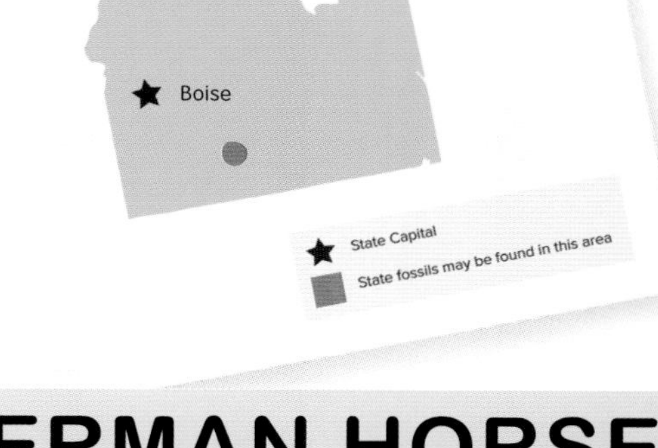

Pliocene epoch to Pleistocene epoch (3.5 million to possibly 11,700 years ago)
The Hagerman Horse was designated the state fossil in 1988.

HAGERMAN HORSE

Equus simplicidens
(pronounced Ek-wis sim-plis-id-ens)

The bones of fossil horses were discovered near Hagerman, Idaho, in 1928. The horses were initially given the name *Plesippus shoshonensis* (pronounced Ples-ih-puss sh-oh-sh-oh-nen-sis) in 1930, before it was discovered to be the same as a fossil horse discovered from Texas named in 1892: *Equus simplicidens*.

While *Equus simplicidens* became the official name, the common name Hagerman Horse is the one everyone uses. The Hagerman Horse was 4 to 5 feet tall and had a body shape similar to a zebra's. It roamed the grasslands of Pliocene North America before dying out some time in the Pleistocene, some 3.4 million years later. Further excavation of the area near Hagerman eventually became the Hagerman Horse Quarry, where several complete skeletons and more than 100 skulls of the Hagerman horse have been found. The quarry became part of Hagerman Fossil Beds National Monument when it was established in 1988.

Hagerman Horse cast at the Arizona Museum of Natural History

Illinois

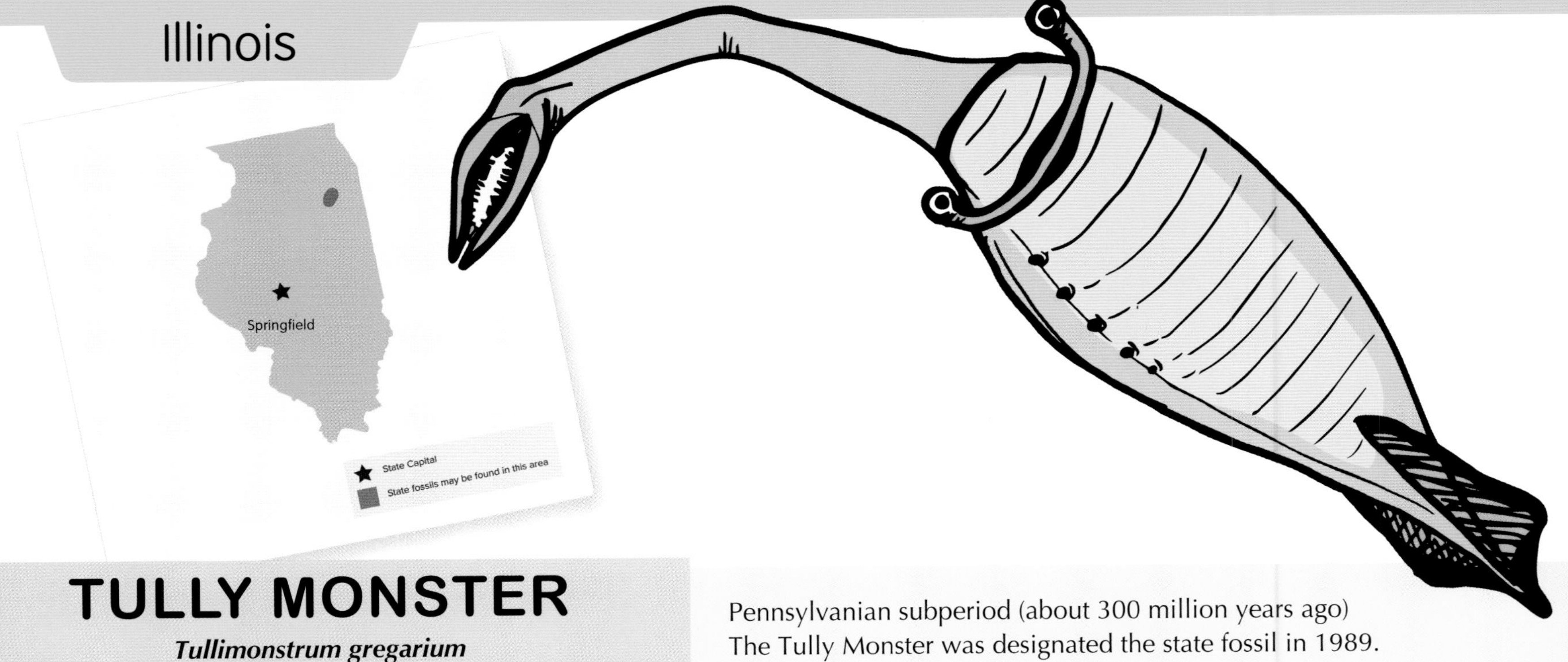

TULLY MONSTER

Tullimonstrum gregarium
(pronounced Tull-ee-mun-strum greg-ah-ree-um)

Pennsylvanian subperiod (about 300 million years ago)
The Tully Monster was designated the state fossil in 1989.

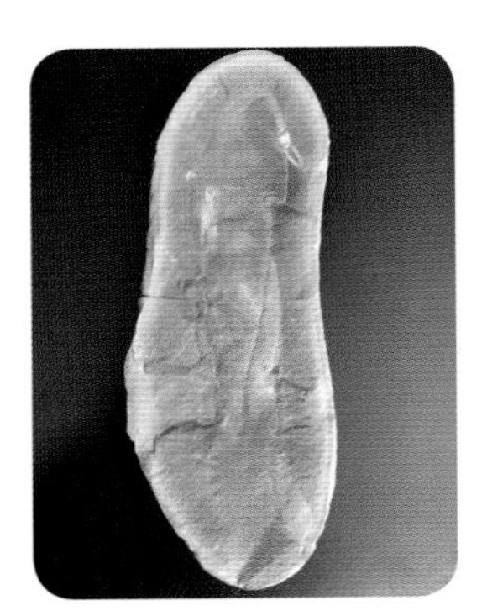

Fossil Tully Monster in a nodule

Francis Tully, a fossil hunter, found a weird-looking fossil in 1955, in the coal strip mines he regularly hunted near Braidwood, Illinois. He took it to the Field Museum of Natural History where puzzled paleontologists called it Tully's Monster, which became the genus Tullimonstrum. The Tully Monster was indeed a monster; it had a trunk like an elephant that ended in a claw with teeth, eyes on a pair of stalks, and a pair of fins near the tail to help it move. It grew up to 14 inches long and today paleontologists are still debating what kind of animal it was. Some say it was an early vertebrate while others say it was an invertebrate. Tully Monsters are occasionally found in nodules in coal mine spoil piles of the famous Mazon Creek fossil deposits of the Francis Creek shale.

Indiana

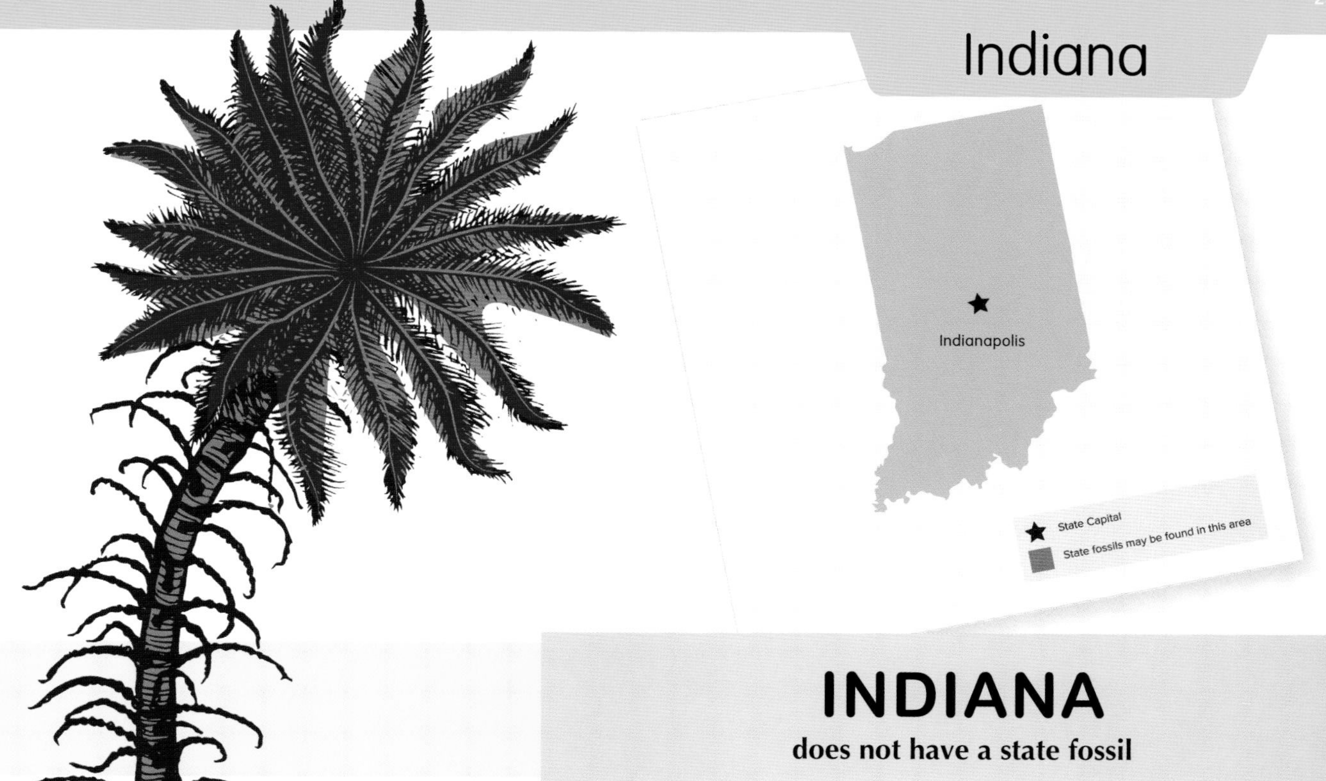

INDIANA

does not have a state fossil

Recommendation for state fossil:

Indiana has a variety of marine fossils of various ages, but most notable are the well-preserved fossil crinoids in the Edwardsville formation near Crawfordsville, Indiana. Crinoids are echinoderms, related to sea stars and sea urchins. Crinoids, also called sea lilies, are filter feeders, and more than 600 species are alive today. The Edwardsville formation is early Mississippian in age, about 350–340 million years old, and within it are dozens of species of well-preserved crinoids. In 2015, there was a bill, or a proposed law, to make the crinoid *Elegantocrinus hemisphaericus* (pronounced El-ig-an-toe-krie-nus hem-is-feh-rik-us) the state fossil, but the bill did not make it far.

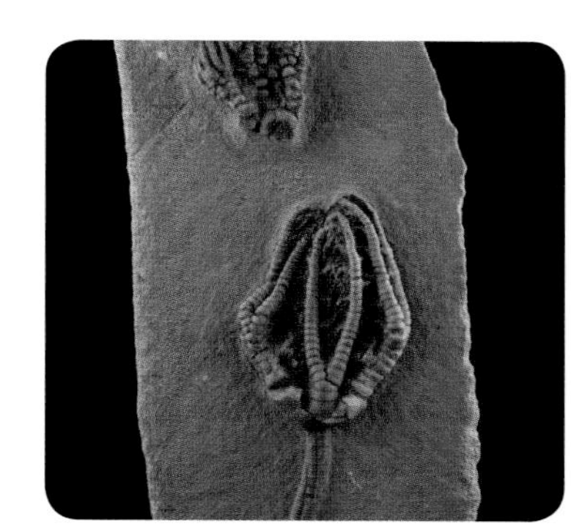

Crinoid fossil from the Edwardsville formation of Indiana

Iowa

IOWA

does not have a state fossil

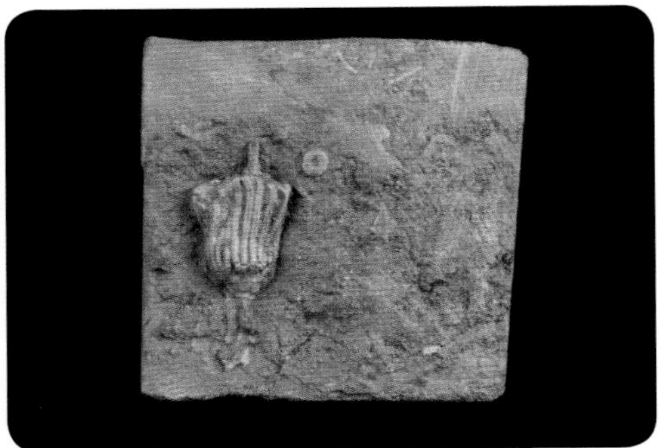

Fossil crinoid *Eretmocrinus tentor* from Iowa

Recommendation for state fossil:

Iowa has a rich variety of fossils including trilobites, brachiopods, corals, gastropods, plants, and mammoths. Many finely preserved crinoids have been found near Le Grand and Gilmore City. Crinoids are echinoderms, related to sea stars and sea urchins, and some species are still alive today. Crinoids often disintegrate after death, leaving behind mostly round stem segments, but many fine complete specimens have been found in Iowa. In 1996, there was a proposed bill to designate the crinoid as the state fossil but the bill went nowhere. The same happened to another crinoid bill in 1999.

Kansas

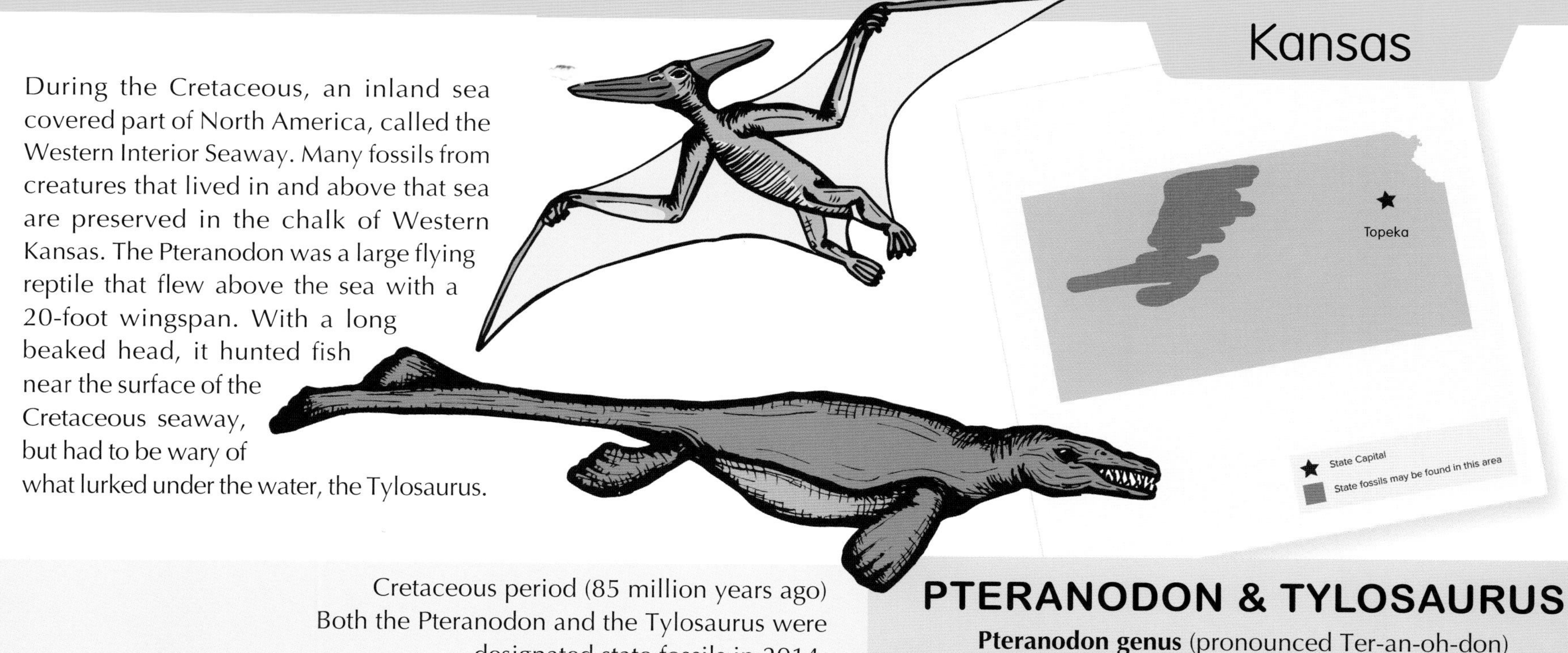

During the Cretaceous, an inland sea covered part of North America, called the Western Interior Seaway. Many fossils from creatures that lived in and above that sea are preserved in the chalk of Western Kansas. The Pteranodon was a large flying reptile that flew above the sea with a 20-foot wingspan. With a long beaked head, it hunted fish near the surface of the Cretaceous seaway, but had to be wary of what lurked under the water, the Tylosaurus.

Cretaceous period (85 million years ago)
Both the Pteranodon and the Tylosaurus were designated state fossils in 2014.

PTERANODON & TYLOSAURUS

Pteranodon genus (pronounced Ter-an-oh-don)
Tylosaurus genus (pronounced Tie-loh-saw-rus)

A giant marine reptile that grew up to 45 feet long, the Tylosaurus swam using four flippers and had a large skull full of sharp teeth. The Tylosaurus was a large predator that ate whatever it could get, including fish, sharks, and other marine reptiles. Extraordinary well-preserved fossils of both have been found throughout the Niobrara chalk of Western Kansas.

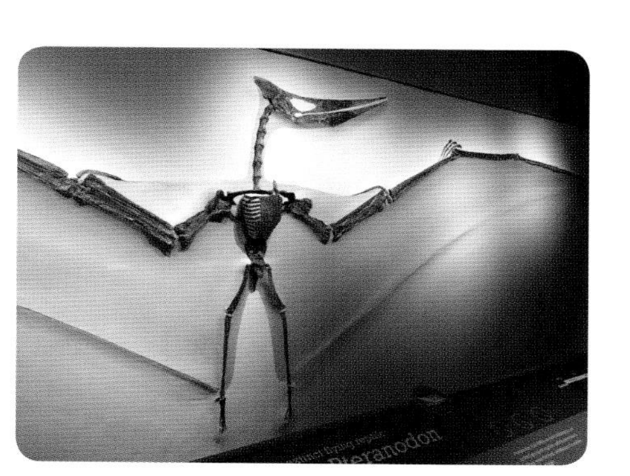
Pteranodon skeleton at the University of Kansas Natural History Museum

Tylosaurus skeleton at the University of Kansas Natural History Museum

Kentucky

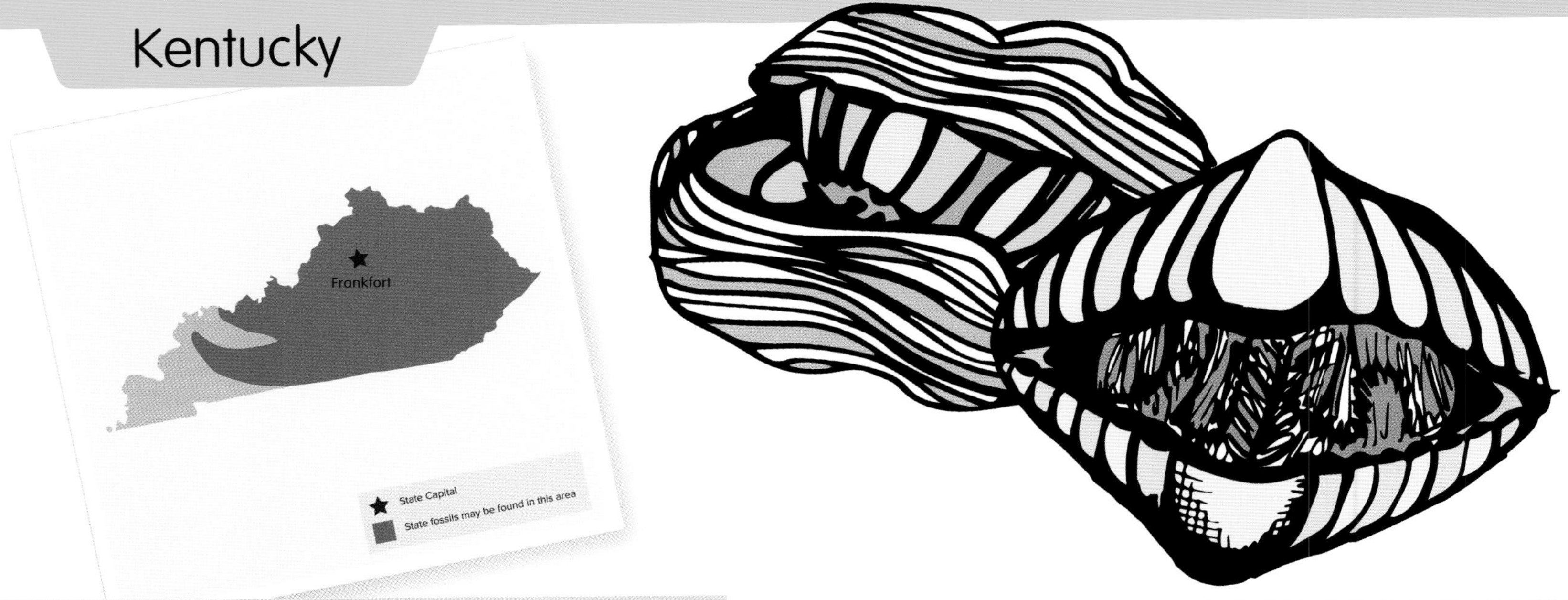

BRACHIOPOD

Brachiopoda phylum
(pronounced Brak-ee-o-poh-dah)

Ordovician period to Pennsylvanian subperiod (485 to 299 million years ago)
The brachiopod was designated the state fossil in 1986.

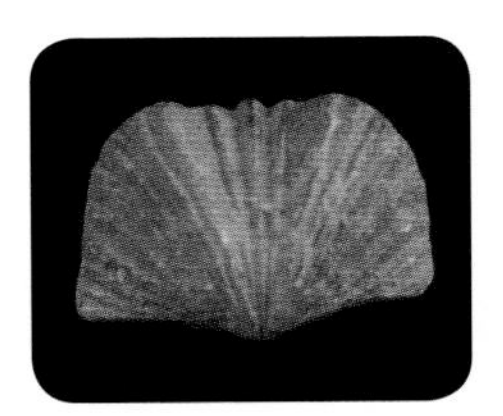

Example of a brachiopod from Kentucky

For hundreds of millions of years, shelled organisms called brachiopods occupied the oceans and were one of the most abundant filter feeding organisms. While brachiopods look similar to clams and oysters, they are an entirely different phylum, Brachiopoda, and not molluscs. Unlike molluscs, each half (or valve as it's called) of the brachiopod shell does not mirror the other, but one is usually larger or slightly different from the other. The larger side of the shell also has a small hole in it where a stalk would have anchored the brachiopod to a rock or substrate. Brachiopods mostly died out during an extinction at the end of the Permian period, which is why molluscs are more common today, but a few hundred species of brachiopods still live in the oceans. Brachiopods are very common fossils in Kentucky, with numerous species being found throughout the state, so the general term brachiopod was designated the state fossil rather than just one species.

Louisiana

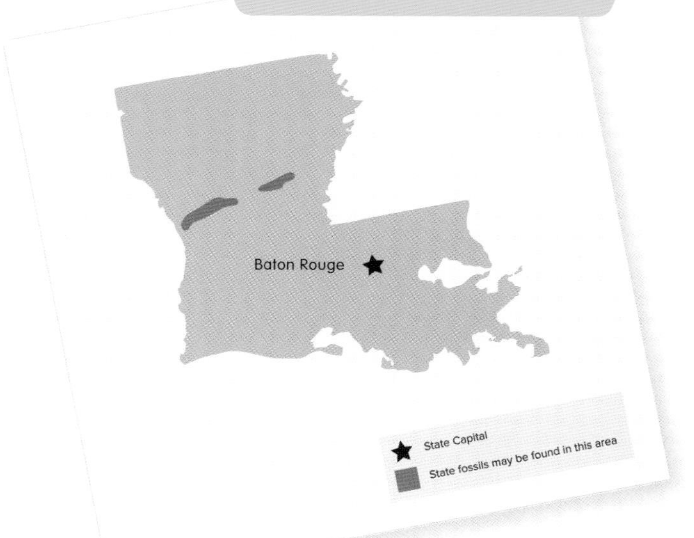

Oligocene epoch (33 to 23 million years ago)
Petrified palmwood was designated the state fossil in 1976.

PETRIFIED PALMWOOD

Palmoxylon genus
(pronounced Pal-m-ah-ks-eh-lon)

Palm trees are uncommon in the fossil record, which is why specimens of petrified palmwood from Louisiana are special. During the Oligocene, palm trees of the Palmoxylon genus grew along the beaches of what was then the coastline, and they were buried and preserved in marine sediments. The palmwood in Louisiana is preserved in silica and comes in a variety of colors, often with an attractive spotted pattern. The spots are the remains of small, round bundles of tissue that once provided the tree with structural support.

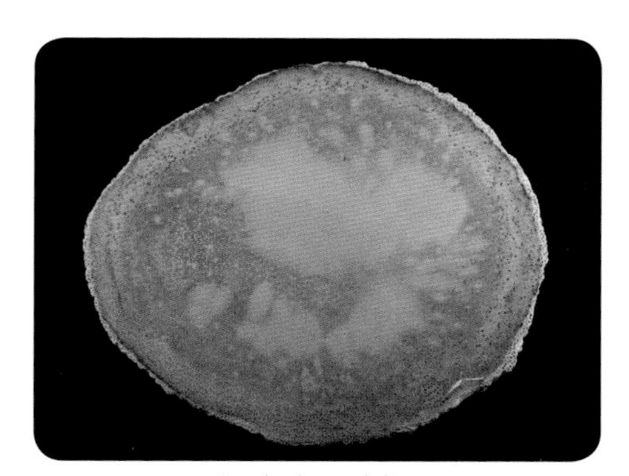

Fossil palmwood slice

Maine

Augusta

State Capital

State fossils may be found in this area

PERTICA

Pertica quadrifaria

(pronounced Per-tik-ah kw-ah-drif-eh-ree-ah)

Middle Devonian period (about 390 million years ago)

The *Pertica quadrifaria* was designated the state fossil in 1985.

Pertica specimen from the Trout Valley formation

An unusual fossil plant was found in the Trout Valley formation of Maine in 1968, and was named *Pertica quadrifaria* in 1972. It was a vascular plant that grew up to six feet tall, with a central stem with four rows of branches. These branches were stacked in a spiral formation around the stem. There were no leaves on these branches, but rather some of the branches ended in forked tips while others ended with bunches of sporangia (spore casing) that helped it reproduce. The Pertica belonged to a group of plants informally called Trimerophytes (pronounced Try-mer-oh-fie-tes) that many scientists believe to be the ancestor group from which most modern plants evolved.

Maryland

Note:
The ***Astrodon johnstoni*** (pronounced As-tro-don john-ston-eye) was designated the official state dinosaur in 1998. It was a long-necked dinosaur first discovered in 1858.

Late Miocene epoch (about 5.3 million years ago)
The Ecphora was designated the state fossil shell in 1994, after a species change from its original designation in 1984.

SEA SNAIL

Ecphora gardnerae gardnerae (Wilson)
(pronounced Eck-for-ah gah-r-d-neh-ray)

Maryland technically has a state fossil shell, but most people consider it to be the state fossil. The *Ecphora gardnerae* was a predatory sea snail that lived in the seas covering parts of Maryland during the late Miocene. It was carnivorous and ate other molluscs by boring holes into their hard shells. The fossils of the Ecphora are delicate, but many have been found along the famous Calvert Cliffs of Maryland that border the Chesapeake Bay. The same snail was previously designated the Maryland state fossil shell in 1984, as *Ecphora quadricostata*, but the species was renamed and designation was changed to reflect that in 1994. What's unusual is that this updated law was very specific about the species name, to the point of including the subspecies name and the name of the scientist who discovered that it was a separate species. The updated law was written *Ecphora gardnerae gardnerae* (Wilson).

Ecphora from the Calvert Cliffs area of Maryland

Massachusetts

DINOSAUR TRACKS

Early Jurassic period (about 200 to 190 million years ago)
Dinosaur tracks were designated the state fossil in 1980.

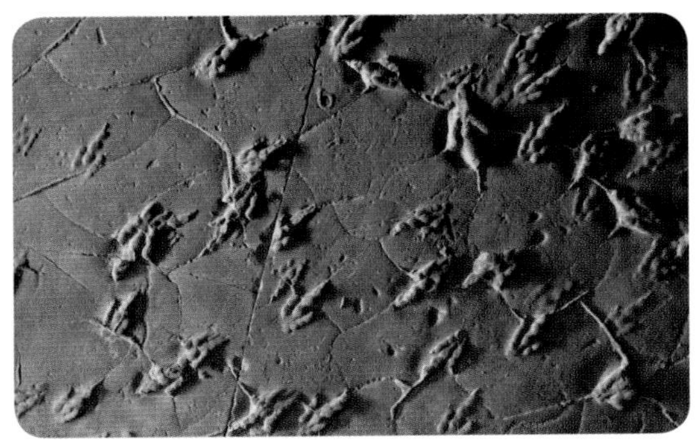

Fossil trackway from the Connecticut River Valley, at the Beneski Museum of Natural History

Massachusetts is famous for numerous dinosaur tracks that have been discovered in the Connecticut River Valley of Western Massachusetts since the early 1800s. Many types of tracks are found in Massachusetts, from small 3- to 5-inch three-toed tracks called Grallator (pronounced Grah-la-tore) that were made by small bipedal theropod dinosaurs, to 11- to 19-inch tracks called Eubrontes (see Connecticut state fossil) that may have been made by the Dilophosaurus. The largest collection of dinosaur tracks from the Connecticut River Valley can be seen at the Beneski Museum of Natural History at Amherst College: it contains more than a thousand trackways composed of thousands of dinosaur footprints.

Michigan

Note:
The **Petoskey** (pronounced Pet-os-kee) stone, a fossilized coral, was designated the state stone in 1965.

Pliocene epoch to Holocene epoch
(4.75 million years to about 10,500 years ago)
The mastodon was designated the state fossil in 2002.

MASTODON

Mammut americanum
(pronounced mah-mut ah-mer-i-can-um)

The mastodon was very similar to elephants and mammoths but not closely related to either. It had a long, low skull with curved tusks, grew up to 10 feet tall, and weighed up to 8 tons. Like the woolly mammoth, the mastodon also had a shaggy coat, but the mastodon had distinctive cusp-shaped teeth and tusks up to 16 feet in length.

Mastodons wandered around Michigan during the Ice Age, and parts of more than 300 mastodons have been found in the state. In 1992, a trail of mastodon footprints was discovered near Saline. Composed of more than 30 footprints measuring up to 20 inches across, it is the longest mastodon trackway in the world.

Mastodon at the Beneski Museum of Natural History

Minnesota

MINNESOTA

Does not have a state fossil

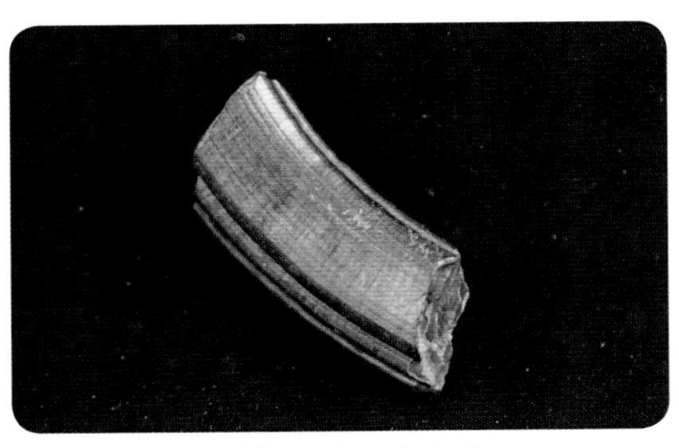

Tooth of the Giant Beaver, Castoroides

Recommendation for state fossil:

Minnesota has a variety of marine fossils and some terrestrial fossils, but one that stands out was a giant beaver. In 1988, a bill was introduced to designate the giant beaver, *Castoroides ohioensis* (pronounced Kas-toh-roy-dess oh-hi-oh-en-sis), as the state fossil of Minnesota. The bill did not pass. The Castoroides lived during the late Pleistocene from 130,000 to 11,700 years ago and grew up to 7 feet long, weighed up to 275 pounds, and had huge 6-inch incisors. Fossils of Castoroides have been found throughout the United States including well-preserved specimens in Minnesota.

Mississippi

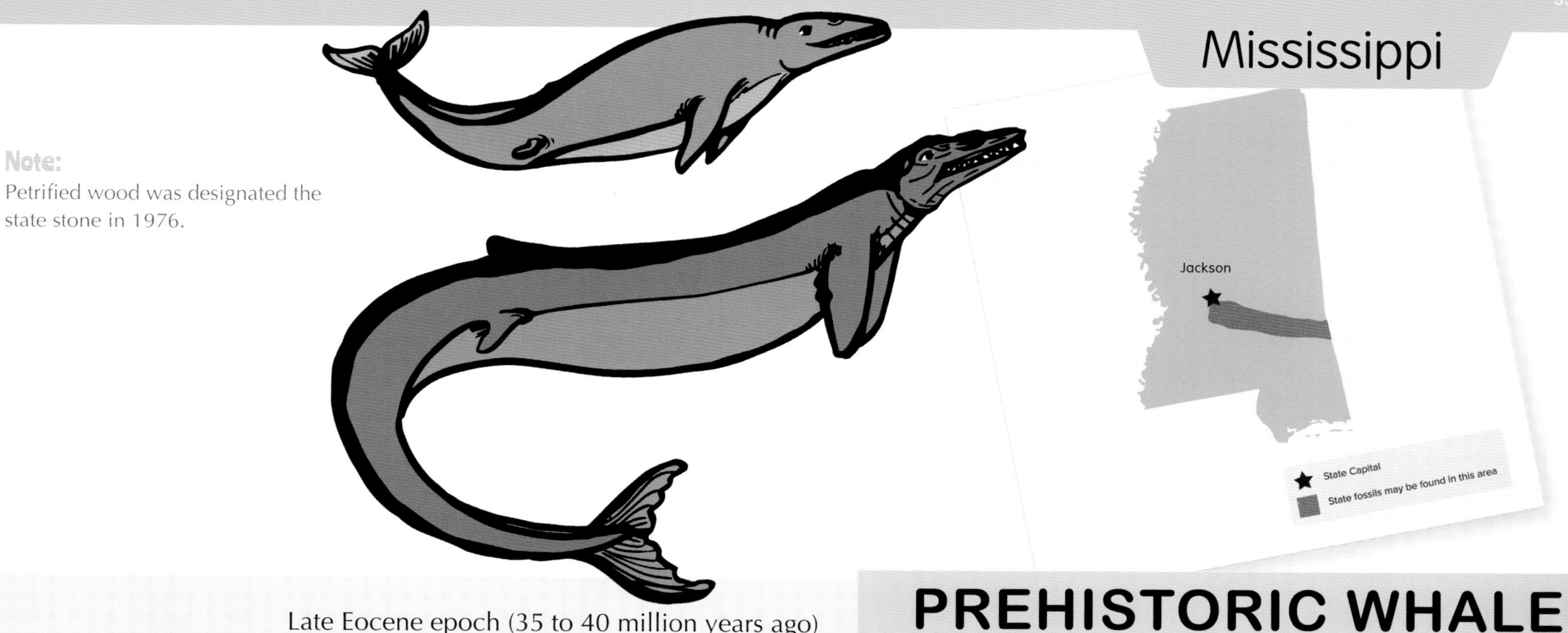

Note:
Petrified wood was designated the state stone in 1976.

Late Eocene epoch (35 to 40 million years ago)
The "prehistoric whale" was designated the state fossil in 1981.

PREHISTORIC WHALE

Basilosaurus cetoides and ***Zygorhiza kochii***
(pronounced Zie-go-rise-ah k-ah-chee-i)

The bones and skeletons of fossil whales have been found in the late Eocene sediments of Mississippi since the 1830s. The two species found in Mississippi are the *Basilosaurus cetoides* and the *Zygorhiza kochii*. The Basilosaurus grew up to 59 feet long and had a 5-foot-long, wedge-shaped head. It also had cone-shaped teeth at the front of the jaws for holding prey and triangular shaped teeth at the back of the skull for slicing. The Zygorhiza grew up to 20 feet long and also had sharp teeth. In 1971, a near-complete Zygorhiza skeleton was discovered in a creek bed near Tinsley, Mississippi. This skeleton helped stir public interest, which eventually led to the resolution designating the prehistoric whale as the state fossil.

Zygorhiza kochii at George Mason University

Missouri

Note:
The ***Hypsibema missouriensis*** (pronounced Hip-so-bee-mah mis-oh-r-ee-eh-n-sis) was designated the state dinosaur in 2004. It was a Hadrosaur from the Late Cretaceous.

CRINOID—SEA LILY

Delocrinus missouriensis
(pronounced Del-oh-cry-n-us mis-oh-r-ee-eh-n-sis)

Pennsylvanian subperiod (323 to 299 million years ago)
The crinoid was designated the state fossil in 1989.

Rare complete *Delocrinus missouriensis* specimen

Crinoids are marine animals that are often called sea lilies and are still around today. They are part of the Echinodermata phylum, which includes sea urchins, sea stars, and sand dollars. Crinoids are filter feeders; they catch particles of food using their feathery arms, which have a system to transport food to the calyx. The calyx has a mouth and digestive system. On most crinoids, the calyx is attached to the sea floor by a long stem. Crinoid fossils can be found in many places in Missouri, with preserved stem segments being the most common part found. The *Delocrinus missouriensis* is a species of crinoid that is rare but can be found in some Pennsylvanian deposits throughout Missouri.

Montana

Upper Cretaceous period (76 million years ago)
The *Maiasaura peeblesorum* was designated the state fossil in 1985.

DUCK-BILLED DINOSAUR

Maiasaura peeblesorum
(pronounced My-ah-saw-rah pee-bo-soh-rum)

The *Maiasaura peeblesorum* was a member of the Hadrosauridae family, also known as duck-billed dinosaurs because their snouts look like the bills (beaks) of ducks. In 1978, fossilized nests were found in Montana along with the remains of juvenile dinosaurs. The presence of juvenile dinosaurs suggested that the adult dinosaurs fed and took care of their young in the nest. This evidence of nurturing resulted in the naming of Maiasaura, meaning "good mother lizard" in Latin. Maiasaura were herbivores and grew up to 30 feet long. They created nesting colonies with nests of up to 40 eggs. Each egg was about 5 to 6 inches long. More than 200 specimens of Maiasaura have been found.

Mounted Maiasaura skeleton

Nebraska

THE MAMMOTH

Mammuthus primigenius, Mammuthus columbi
(pronounced Mah-muh-thus coh-lumb-eye)

Pleistocene epoch (1.1 million years to about 11,700 years ago)
The mammoth was designated the state fossil in 1967.

Close up of Columbian mammoth skull at the Beneski Museum of Natural History

Two varieties of mammoths roamed Nebraska during the Ice Age and their remains have been found in every county of the state. One was the woolly mammoth, *Mammuthus primigenius*, which grew up to 11 feet tall and weighed up to 6 tons. The woolly mammoth was covered in a coat of shaggy, coarse hair. The other was the Columbian mammoth, *Mammuthus columbi*, which grew up to 13 feet tall and weighed up to 10 tons. The Columbian mammoth was bigger than the woolly mammoth and tended to live in warmer areas. In 1922, a rancher and his wife in Lincoln County noticed their chickens were pecking at something unusual. This turned out to be the bones of the largest Columbian mammoth in the world, which people nicknamed Archie. The Imperial mammoth was once considered to be another mammoth of Nebraska, but recent study has determined Imperial mammoths were actually Columbian mammoths.

Nevada

Carson City

State Capital

State fossils may be found in this area

Triassic period (about 215 million years ago)
The Ichthyosaur was designated the state fossil in 1977, and was amended to be the Shonisaurus in 1989.

GIANT ICHTHYOSAUR

Shonisaurus popularis
(pronounced Shoh-n-ih-saw-rus pop-yoo-lah-ris)

In 1928, a paleontologist first recognized some giant fossil bones found in the Shoshone Mountains of Nevada as those of a giant Ichthyosaur (pronounced ik-thee-oh-soar). Ichthyosaurs were marine reptiles that had four flippers, a tail, and long skull with sharp teeth. They were fast swimmers and surfaced to breathe air. Further excavations at the site in the Shoshone Mountains revealed the remains of 37 individual giant Ichthyosaurs that died together. The fossils indicated these Ichthyosaurs grew up to 49 feet long and had vertebrae up to a foot across. These giants were named Shonisaurus to honor the mountains where they were discovered. Many of the specimens are on display in the ground where they were found at Berlin-Ichthyosaur State Historic Park. Some of the remains and a replica of a whole one can be seen at the Nevada State Museum in Las Vegas.

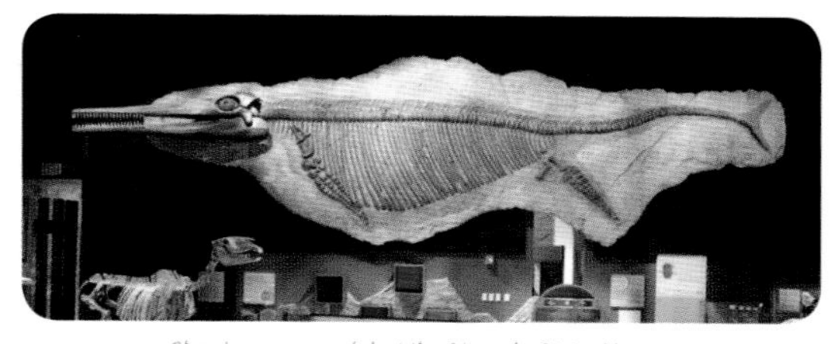

Shonisaurus model at the Nevada State Museum

NEW HAMPSHIRE

Does not have a state fossil

Mastodon tooth from New Hampshire

Recommendation for state fossil:

New Hampshire is not known for many fossils because most of the state has igneous and metamorphic rocks that don't preserve fossils. What few fossils have been discovered include some Silurian period marine fossils found near Littleton, and rare mammoth and mastodon teeth. In 2013, a class of third-graders and their teacher from Bradford, New Hampshire, began a quest to have a state fossil designated. They chose the mastodon because only one other state (see Michigan state fossil) had it as a state fossil and a well-preserved tooth had recently been found off the coast. In 2015, the students introduced the bill to the state legislature. Sadly, it was not passed, but they can try again in the future.

New Jersey

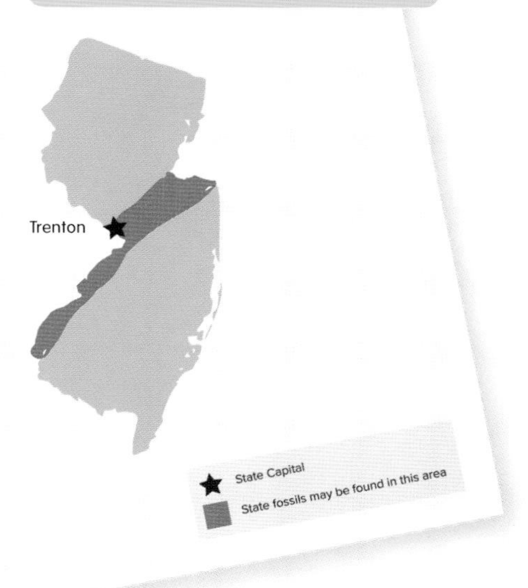

Cretaceous period (78 to 80 million years ago)
The *Hadrosaurus foulkii* was designated the state dinosaur in 1991.

DUCK-BILLED DINOSAUR

Hadrosaurus foulkii
(pronounced Had-roh-saw-rus foul-kee-eye)

New Jersey does not have a state fossil, but it does have a much-loved state dinosaur. Dinosaur bones were excavated from a marl pit in Haddonfield, New Jersey, in 1858, and became the most complete unearthed dinosaur in the world at the time. It was named *Hadrosaurus foulkii* after William Parker Foulke, who helped dig it out. The Hadrosaurus was a duck-billed dinosaur that stood 10 feet tall and measured more than 23 feet long. It was an herbivore with hundreds of small teeth for eating vegetation. In 1868, the bones were put together to become the first assembled dinosaur skeleton on display in the world. It attracted thousands of visitors to its home at the Academy of Natural Sciences in Philadelphia. The site where it was discovered became a National Historic Landmark in 1994.

Hadrosaurus foulkii tooth from New Jersey

New Mexico

COELOPHYSIS

Coelophysis bauri
(pronounced See-loh-f-ie-sis bower-eye)

Late Triassic period (about 205 to 210 million years ago)
The Coelophysis was designated the state fossil in 1981.

Coelophysis skeleton from the Denver Museum of Nature & Science

The Coelophysis is one of the earliest theropod dinosaurs and one of the most well-known. It grew up to 9 feet long and only weighed up to 44 pounds. They had small sharp teeth, indicating they were predators, and likely fed on smaller animals or may have hunted in packs to take down larger prey. Coelophysis are well-known because hundreds of intact specimens have been found at Ghost Ranch in New Mexico, often in massive clumps with a few hundred individuals packed together. It is still debated how all the Coelophysis found at Ghost Ranch died, but it is believed that a flood was responsible for depositing and burying the remains together.

New York

Late Silurian period (423 to 418 million years old)
The *Eurypterid remipes* was designated the state fossil in 1984.

SEA SCORPION

Eurypterus remipes
(pronounced Yoo-rip-tah-rus rem-i-pees)

Eurypterids, also known as Sea Scorpions, were arthropods that lived in the oceans from the Ordovician to the Permian. They belonged in the subphylum Chelicerata, which includes horseshoe crabs, spiders, and scorpions. Some species grew up to 8 feet long and many were armed with claws filled with sharp teeth. The *Eurypterus remipes* is the most common type of Eurypterid found and averaged 5 to 8 inches long, but could grow up to 16 inches long. They had two large paddles for swimming and a long tail called a telson. Fossils of the Eurypterus genus were first discovered in New York in 1818, and many have been found in the Bertie formation throughout New York.

Fossil *Eurypterus remipes* from New York

North Carolina

MEGALODON SHARK TOOTH

C. megalodon
(pronounced Meg-ah-loh-don)

Miocene epoch to Pliocene epoch (23 to 2.6 million years old)
The fossilized teeth of the Megalodon were designated the state fossil in 2013.

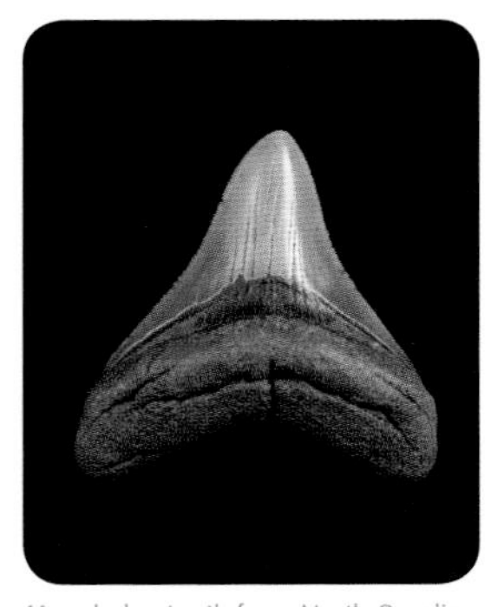
Megalodon tooth from North Carolina

The Megalodon was the largest shark that ever existed. It could have grown up to 59 feet long or even bigger, but no one knows for sure because usually only fossils of its teeth are found. The rest of the Megalodon was cartilage, which rarely fossilizes. It had rows of sharp teeth measuring up to 7 inches long, and these fossilized teeth have been found all over the world. Its mass could have been more than 100 tons, and to grow that big it fed on whales, dolphins, and other marine creatures. Fossil whale bones have been found with bite marks that match those that would have been made by the Megalodon. It is still debated what genus Megalodon belongs to, usually Carcharodon or Carcharocles, which is why it is most often referred to as *C. megalodon*. The teeth of the Megalodon can be found in rivers and seashores throughout Eastern North Carolina.

North Dakota

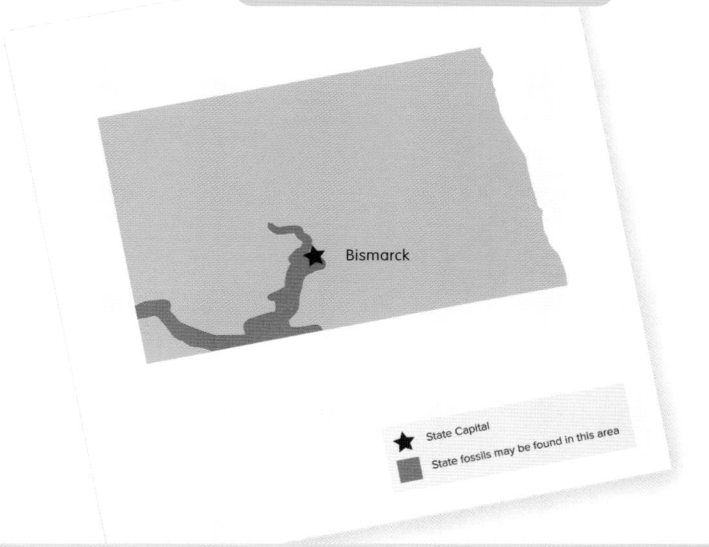

Paleocene epoch (65 to 60 million years ago)
Teredo petrified wood was designated the state fossil in 1967.

SHIPWORM-BORED PETRIFIED WOOD

Teredo petrified wood (pronounced Teh-ray-do)

North Dakota was partially covered by an inland sea called the Cannonball Sea during the Paleocene. Often, trees would fall and get washed out to sea, where Teredo shipworms would bore into them. Shipworms are actually a type of small marine clam. The Teredo shipworms created holes, giving the wood the appearance of Swiss cheese. The hole-ridden wood fossilized to become the attractive Teredo petrified wood, found in many places in North Dakota. Shipworms are still around and continue to make holes in driftwood and wooden ships.

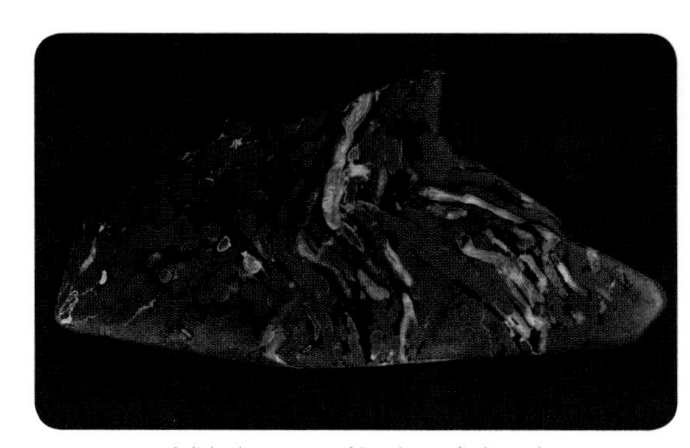

Polished specimen of Teredo petrified wood

Ohio

ISOTELUS TRILOBITE

Isotelus genus
(pronounced Eye-so-tee-lus)

Middle and upper Ordovician period (470 to 444 million years ago)
The Isotelus was designated the state invertebrate fossil in 1985.

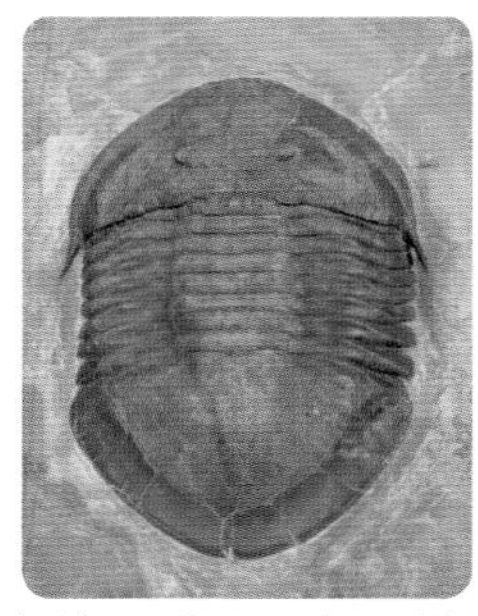
Isotelus maximus specimen at the Karl E. Limper Geology Museum

Trilobites were marine arthropods that thrived in the oceans for millions of years during the Paleozoic. These hard-shelled, segmented creatures walked on many legs on the ocean floor. A trilobite had three sections: a cephalon (head), the thorax (segmented body), and a pygidium (tail). The name trilobite means "three lobes" and does not refer to those sections, but rather to how a trilobite can be divided into three lobes: the left, middle, and right. Isotelus was a genus of trilobite that can be found in Ordovician deposits throughout Southwest Ohio, and some species reached large sizes for trilobites. In 1919, a huge trilobite measuring 14.5 inches was discovered near Dayton, Ohio. That specimen, an *Isotelus brachycephalus*, was displated at the Smithsonian Institution. That find and others inspired Ohio to designate the genus Isotelus as the state fossil (or technically the state invertebrate fossil, though there is no official state vertebrate fossil).

Oklahoma

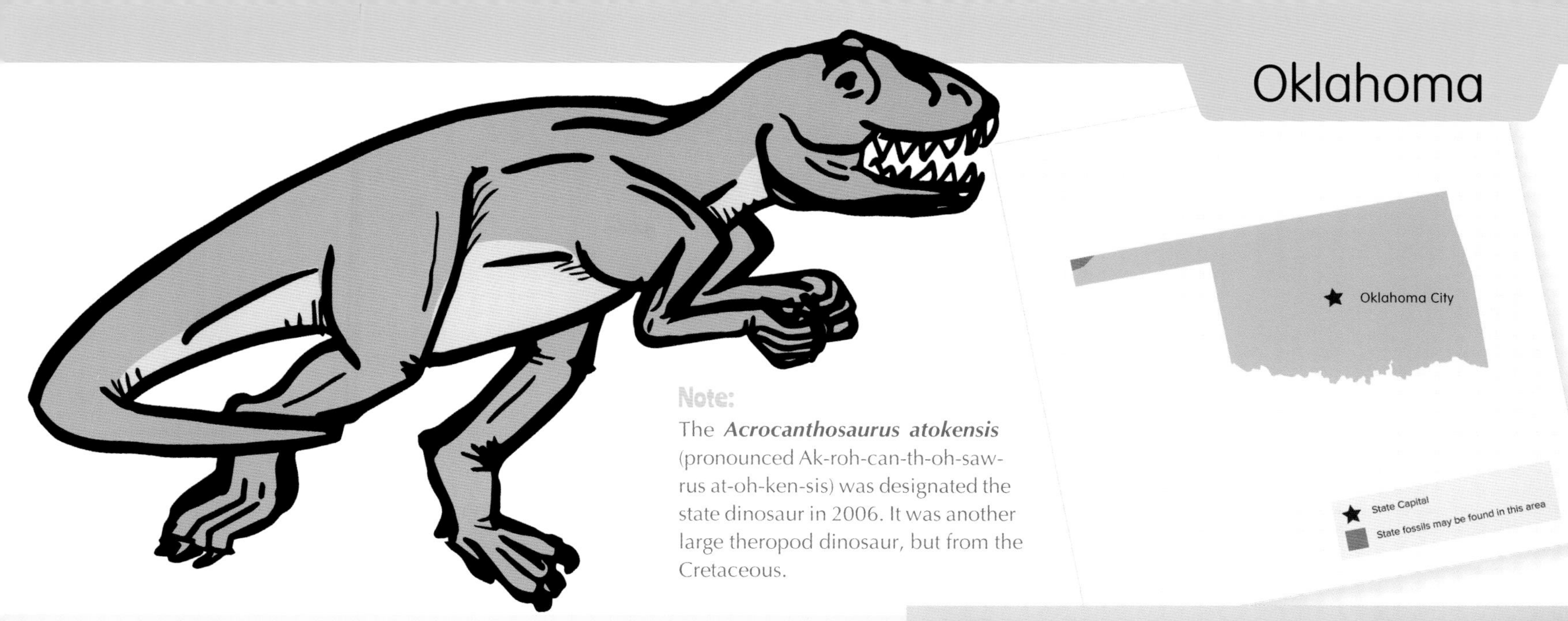

Note:
The ***Acrocanthosaurus atokensis*** (pronounced Ak-roh-can-th-oh-saw-rus at-oh-ken-sis) was designated the state dinosaur in 2006. It was another large theropod dinosaur, but from the Cretaceous.

SAUROPHAGANAX DINOSAUR

Saurophaganax maximus
(pronounced Saw-ro-fa-ga-nax max-i-mus)

Late Jurassic period (156–146 million years ago)
The Saurophaganax was designated the state fossil in 2000.

The *Saurophaganax maximus* was a large theropod dinosaur and one of the larger predators of the Late Jurassic. Fossil fragments of the Saurophaganax were first discovered in the Morrison formation of Western Oklahoma in 1931. These fossils eventually showed the dinosaur could grow up to 43 feet long and weigh up to 3 tons. The Saurophaganax was similar to the Allosaurus, and some believe that it was actually just a very large Allosaurus (see Utah state fossil). With large clawed hands and sharp teeth, the Saurophaganax likely preyed on other dinosaurs of the Morrison formation, such as the Stegosaurus (see Colorado state fossil).

Drawing of a Saurophaganax skull.

Oregon

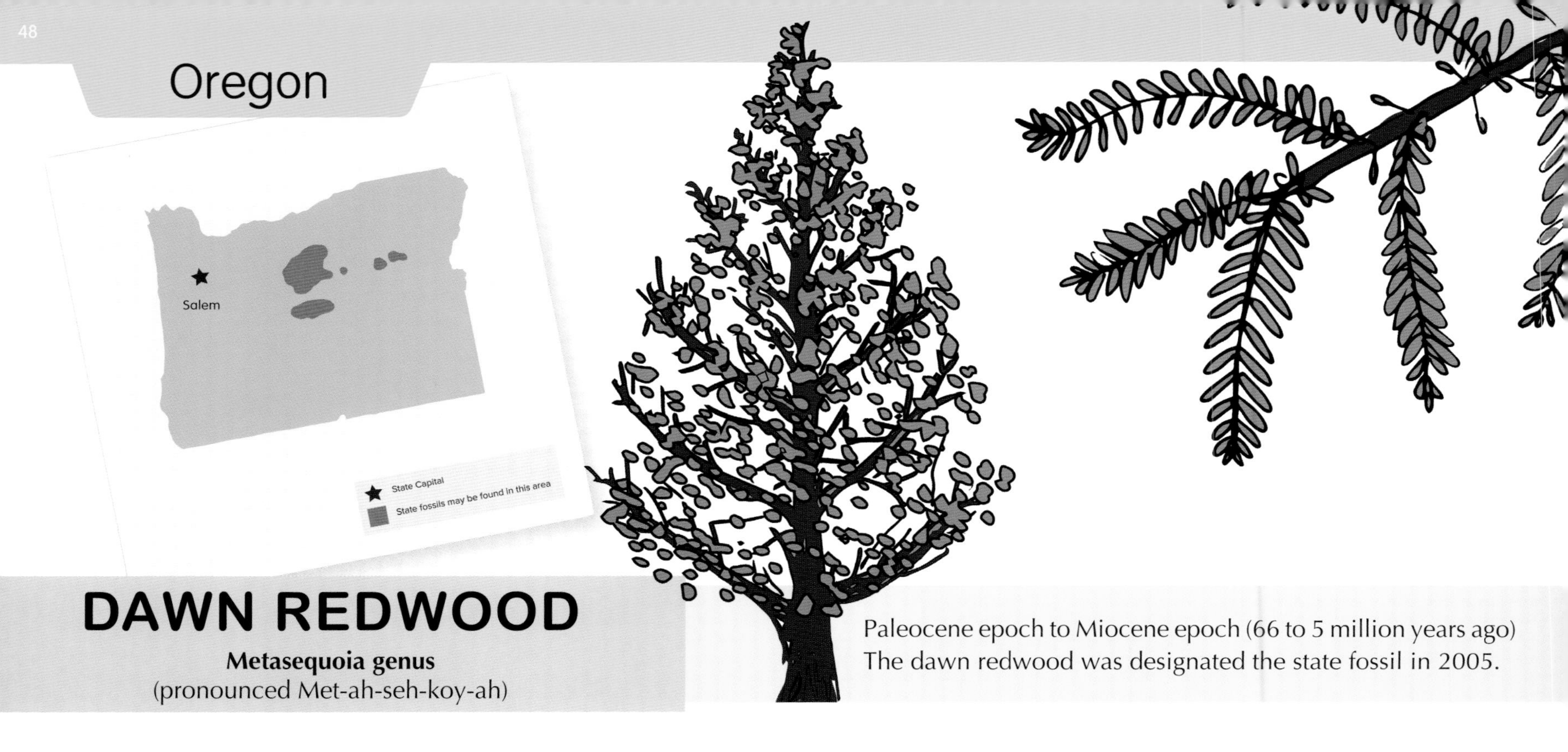

DAWN REDWOOD

Metasequoia genus
(pronounced Met-ah-seh-koy-ah)

Paleocene epoch to Miocene epoch (66 to 5 million years ago)
The dawn redwood was designated the state fossil in 2005.

Metasequoia specimen from nearby Washington State

The Dawn redwood, or Metasequoia, was first described as a fossil in 1941. It was thought to be extinct, but then a grove of living dawn redwoods was discovered in China. It is a fast-growing redwood conifer tree, but unlike other redwoods such as the sequoia, the dawn redwood is deciduous and will shed its leaves in autumn. The tree can grow up to 200 feet tall and 6 feet wide. Well-preserved fossilized leaves can be found throughout Eocene and Miocene deposits in Oregon. After their rediscovery in China, living dawn redwoods were planted around the world and can now be found living in Oregon as well.

Pennsylvania

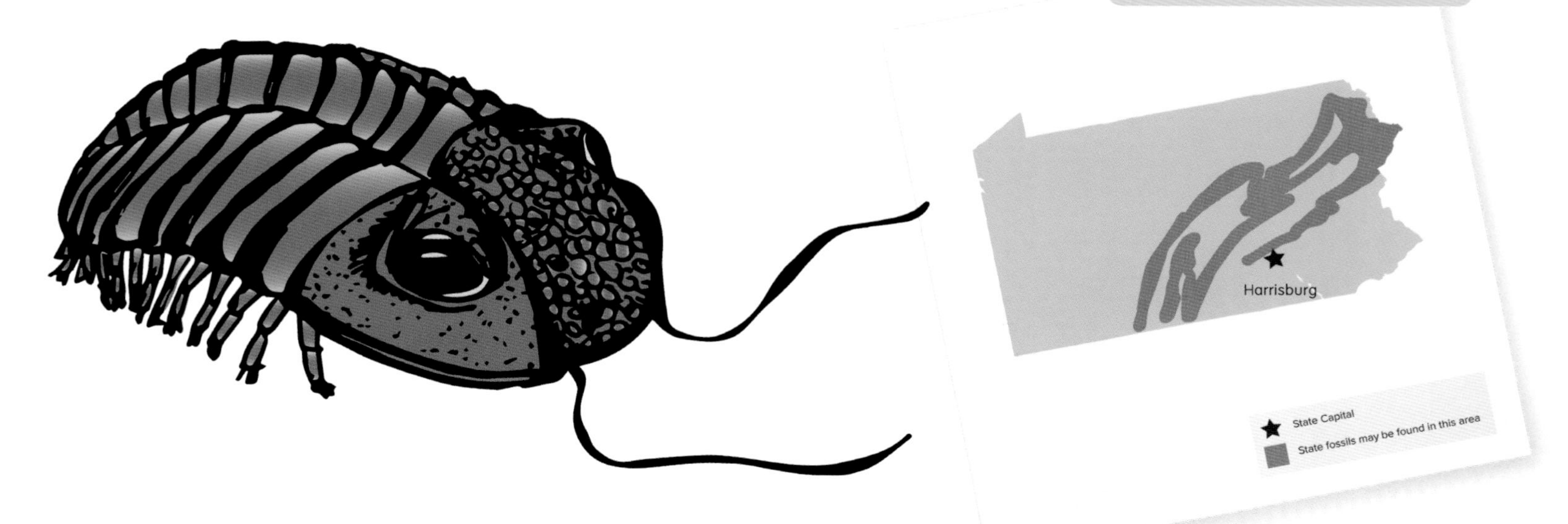

Middle-Devonian period (about 400 to 385 million years ago)
The *Phacops rana* was designated the state fossil in 1988.

PHACOPS TRILOBITE

Phacops rana
(pronounced Fae-cops rah-nah)

Trilobites were arthropods that lived in the oceans from the early Cambrian period to the end of the Permian period. They were highly successful organisms that had hard, segmented exoskeletons and walked on many legs. The *Phacops rana* is a trilobite species named for having eyes that looked like those of frogs, but the Phacops actually had compound eyes with dozens of lenses, like a fly's. The *Phacops rana* is found throughout Pennsylvania in middle Devonian marine deposits. Like many trilobite species, it could roll into a ball for protection and is often found fossilized in that position.

Phacops rana trilobite showing frog-like features

RHODE ISLAND

Does not have a state fossil

Recommendation for state fossil:

Some areas in Rhode Island have deposits from the Carboniferous period, and within these deposits are numerous plant and arthropod fossils. Well-preserved plants including *Pecopteris* (pronounced Peh-cop-ter-is) ferns and horsetail-like *Annularia* (pronounced Ann-u-lar-ee-a) are found in the Carboniferous deposits. The wings of fossil cockroaches and other insects are rare but have been found. One unusual fossil is the *Anthracomartus woodruffi* (pronounced An-th-rah-c-oh-mar-tus wood-ruff-i), which was found near Providence and described in 1893. It is the first arachnid from the Carboniferous discovered in the Eastern United States. Some poorly preserved trilobites are also known from the state.

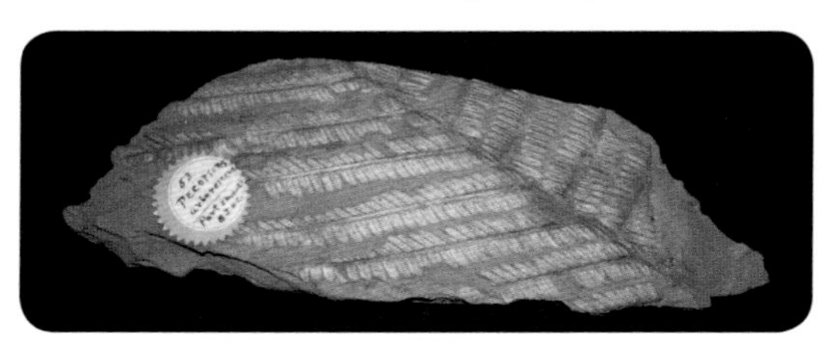

Pecopteris fern specimen from Rhode Island

South Carolina

Pleistocene epoch (1.1 million to 11,700 years ago)
The Columbian mammoth was designated the state fossil in 2014.

COLUMBIAN MAMMOTH

Mammuthus columbi
(pronounced Mah-muh-thus coh-lumb-eye)

In 1725, a naturalist named Mark Catesby visited a plantation in South Carolina and noted that slaves had dug up several large teeth, which they had identified as being from an elephant. Though Columbian mammoths were not formally described until 1857, this identification by slaves was the first technical identification of an American fossil vertebrate, making the Columbian mammoth and South Carolina important milestones in United States fossil history. The Columbian mammoth grew up to 13 feet tall and weighed up to 10 tons. It ate vegetation and had molars that were similar in appearance to those of modern elephants.

Columbian mammoth at the Beneski Museum of Natural History

South Dakota

Pierre

State Capital

State fossils may be found in this area

TRICERATOPS

Triceratops genus

(pronounced Try-sara-tops)

Late Cretaceous period (68 to 66 million years ago)

The Triceratops was designated the state fossil in 1988.

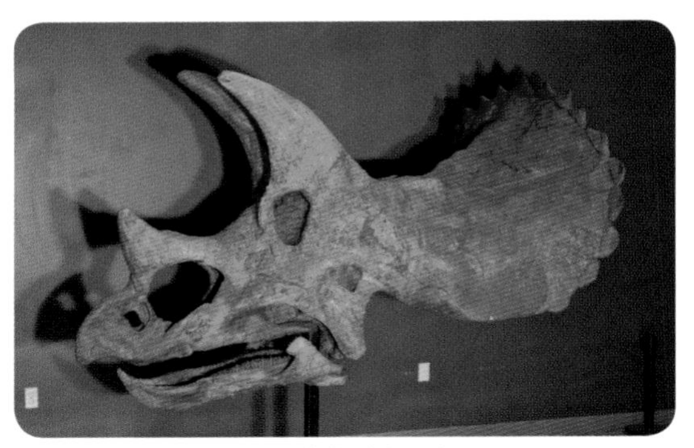

Triceratops skull with large horns

The Triceratops was a large plant-eating dinosaur that walked on all fours and had a distinctive head with three horns and a bony frill. The horns of the Triceratops can measure over 3 feet long. The first Triceratops horns that were discovered were initially believed to be from an unusual fossil bison. The Triceratops roamed throughout much of Western North America and became extinct at the end of the Cretaceous. It weighed up to 26,000 pounds and grew up to 29 feet long. The Triceratops had a hard beak used to tear vegetation and had many replacement teeth because it chewed so much. Many Triceratops fossils have been found in the Cretaceous deposits of South Dakota, along with other famous dinosaurs such as the Tyrannosaurus.

Tennessee

"PTERO" THE BIVALVE

Pterotrigonia thoracica

(pronounced Tero-try-goh-nee-ah thor-a-sik-ah)

Cretaceous period (70 million years ago)

The *Pterotrigona thoracica* was designated the state fossil in 1998.

The *Pterotrigonia thoracica*, nicknamed Ptero by fossil hunters, was a Cretaceous marine bivalve from the Trigoniidae (pronounced Try-gon-ee-i-day) family. The Trigoniidae have very elaborate ornamentation and are suspension feeders that live on the sea floor. The Trigoniidae were thought to have gone extinct at the end of the Cretaceous, but in 1802, a living member of the Trigoniidae family was discovered near Australia. The *Pterotrigonia thoracica* is extinct and its fossils are found in Cretaceous deposits in Western Tennessee. The fossils are numerous in some locations, but they tend to be very fragile, so excavating a complete specimen can be difficult.

A nice *Pterotrigonia thoracica* specimen

Texas

Note:
Petrified palmwood was designated the state stone in 1969 (see Louisiana state fossil).

PALUXYSAURUS

Paluxysaurus jonesi
(pronounced Pal-uk-see-saw-rus jones-eye)

Cretaceous period (115 to 110 million years ago)
The Paluxysaurus was designated the state dinosaur in 2009, after a name change from when it was originally designated in 1997.

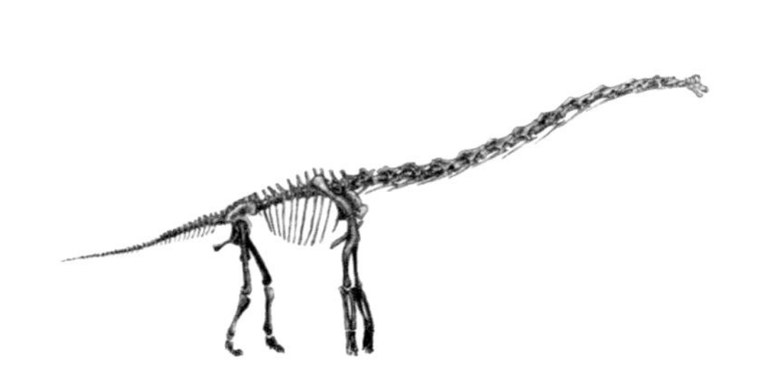

Drawing of a Paluxysaurus (Sauroposeidon) skeleton

Texas does not have a state fossil, but does have a state dinosaur. The Paluxysaurus, a sauropod dinosaur, roamed Texas in the early Cretaceous. Originally it was identified as Pleurocoelus, that was designated the state dinosaur in 1997. However, it was re-identified as Paluxysaurus in 2007, and the state dinosaur name was fixed in 2009. Paluxysaurus bones have been found in North Central Texas, and large fossil tracks in the region have also been attributed to it. It grew to over 60 feet long and had an unusually long neck that distinguished it from other Sauropod dinosaurs of the time. In 2012, researchers re-identified Paluxysaurus as *Sauroposeiden proteles* (pronounced Saw-roh-poh-sie-don pro-te-les), so maybe the state dinosaur name will change again.

Utah

Note:
The **Utahraptor** (pronounced Yoo-ta-rap-tor) was designated the official state dinosaur in 2018. It was a theropod dinosaur that grew up to 19 feet long and had curved claws on its feet measuring up to 9 inches long.

Late Jurassic period (156 to 146 million years ago)
The Allosaurus was designated the state fossil in 1988.

ALLOSAURUS

Allosaurus genus
(pronounced Al-oh-saw-rus)

The Allosaurus was a carnivorous theropod dinosaur, a top predator of the Late Jurassic age Morrison formation. It grew up to 39 feet long and was armed with a mouth filled with dozens of sharp teeth, and claws on its hands that measured up to 7 inches long. *Allosaurus fragilis* was the most common Allosaurus species found. At the Cleveland-Lloyd Dinosaur Quarry in Emory County, Utah, the bones of 46 individual *Allosaurus fragilis* have been found. Allosaurus fed on other dinosaurs, including the Stegosaurus (see Colorado state fossil), as evidenced by bite marks matching Allosaurus mouths found on Stegosaurus bones and a puncture wound on an Allosaurus vertebra that matches a Stegosaurus tail spike.

Allosaurus from the Cleveland-Lloyd quarry on display at Princeton University

Vermont

The White Whale Skeleton was originally designated the state fossil in 1993. In 2014, the white whale skeleton was designated the state marine fossil, and the Mount Holly Mammoth Tooth and Tusk was designated the state terrestrial fossil.

WHITE WHALE SKELETON AND MOUNT HOLLY MAMMOTH TOOTH AND TUSK

Delphinapterus leucas & ***Mammuthus primigenius*** (pronounced Del-fin-ap-terr-us lee-yoo-k-ah-s)

Pleistocene to Holocene epoch (about 12000-11000 years ago)

In 1849, some unusual bones were unearthed near Charlotte, Vermont. These turned out to be from a beluga (white whale), a species still alive today. It had swum into the Champlain Sea, which existed during the late Pleistocene. This skeleton was displayed at the University of Vermont's Perkins Geology Museum and designated Vermont's state fossil in 1993. It was redesignated the state marine fossil in 2014, when the Mount Holly mammoth tooth and tusk was designated the state terrestrial fossil. In 1848, railroad workers near Mount Holly, Vermont, discovered a large tooth and a tusk. These turned out to be from a woolly mammoth and are displayed at the Mount Holly Community Historical Museum.

The "White Whale" skeleton at the Perkins Geology Museum

Mammoth tusk at the Mount Holly Community Historical Museum

Virginia

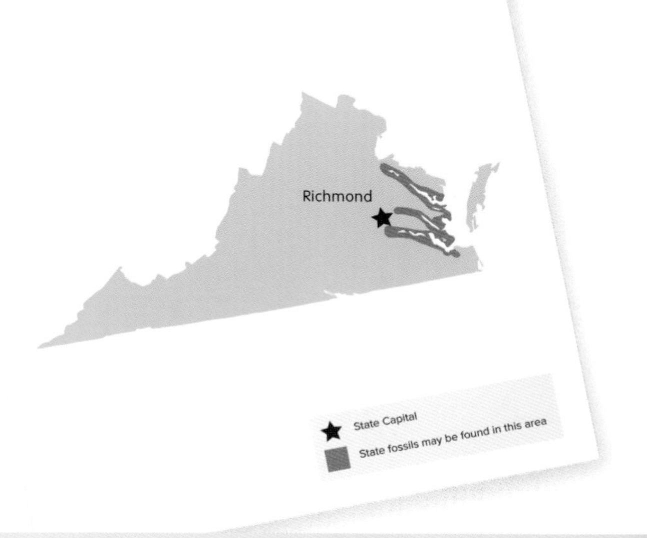

Pliocene epoch (4.5 to 4.3 million years ago)
The Chesapecten was designated the state fossil in 1993.

SCALLOP

Chesapecten jeffersonius
(pronounced Ch-eh-s-ah-pek-ten jef-er-son-ee-us)

The *Chesapecten jeffersonius* was a large scallop that lived along the shallow seas of the Mid-Atlantic coast during the Pliocene. A specimen was described in a scientific publication in 1687, making it the first fossil from the Americas to be scientifically drawn and published. It was named *Pecten jeffersonius* in 1824, to honor former Virginia governor and president Thomas Jefferson. The genus was renamed Chesapecten in 1975, after the Chesapeake Bay area where many of the genus are found. *Chesapecten jeffersonius* only existed for a short time and is used as an index fossil, but can be found washed into rivers and shores throughout Eastern Virginia.

Chesapecten specimen from Virginia

Washington

Note:
Petrified wood was designated the state gem in 1975.

COLUMBIAN MAMMOTH

Mammuthus columbi
(pronounced Mah-muh-thus coh-lumb-eye)

Pleistocene epoch (1.1 million to 11,700 years ago)
The Columbian mammoth was designated the state fossil in 1998.

Columbian Mammoth skeleton at the Beneski Museum of Natural History

The Columbian mammoth roamed throughout Washington State and its fossils can be found in Pleistocene deposits throughout the state. Fossils of its distinctive molars and pieces of its tusks are found most often, but near-complete specimens have also been found. The Columbian mammoth was taller than the woolly mammoth and lived in warmer climates. In 1886, the Chicago Academy of Science assembled and restored a Columbian mammoth skeleton from bones found near Spokane, Washington; it stood 13 feet high and is believed to be the first mounted mammoth skeleton in North America.

West Virginia

Charleston

State Capital

State fossils may be found in this area

Note:
The fossil coral **Lithostrotionella** (pronounced Lith-oh-str-oh-shun-ell-ah) was designated the state gemstone in 1990.

JEFFERSON'S GROUND SLOTH

Megalonyx jeffersonii
(pronounced Meg-ah-lah-niks jef-er-soh-nee-i)

Pleistocene epoch (2.5 million to 11,700 years ago)
The Jefferson's Ground Sloth was designated the state fossil in 2008.

In 1796, three big claws and some bones from a cave in what is now West Virginia were sent to Thomas Jefferson. He initially identified them as belonging to a large lion, calling it Megalonyx, which means "giant claw." Jefferson later found an article about similar fossils in South America and decided that Megalonyx was probably not a large lion. It was properly identified as a sloth later and was named to honor Jefferson in 1822. The *Megalonyx jeffersonii* was huge, measuring up to 10 feet long and weighing up to 2,400 pounds. It had very large claws but ate a diet of plants. It roamed throughout much of North America during the late Pleistocene.

Megalonyx jeffersonii claw from New Jersey at the New Jersey State Museum

Wisconsin

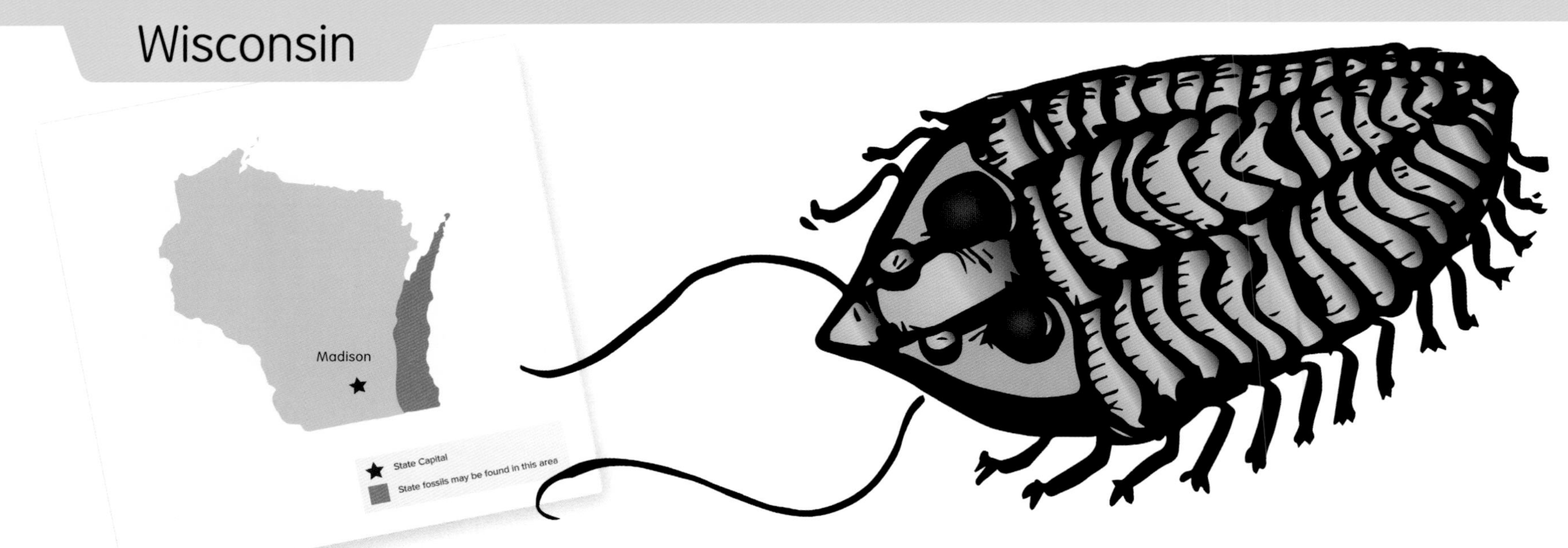

CALYMENE TRILOBITE

Calymene celebra
(pronounced Kal-ih-meen sell-eb-rah)

Silurian epoch (430 to 423 million years ago)
The *Calymene celebra* trilobite was designated the state fossil in 1986.

Calymene Celebra trilobite from Wisconsin

Trilobites lived in the ocean during the Paleozoic. They were marine arthropods that had hard shells, and their fossils can be found throughout the world. One of the most popular trilobites in Wisconsin is the *Calymene celebra*, which can be found throughout several Midwest states. They measure about 2 inches long and are often found as casts and molds in Silurian rock in Eastern Wisconsin. Like many trilobites, the *Calymene celebra* can roll into a ball for protection and occasionally specimens are found in that shape.

Wyoming

Note:
Wyoming passed a statute that a state dinosaur was to be elected. The Triceratops won the election and was designated the state dinosaur in 1994.

FOSSIL FISH

Knightia genus
(pronounced Nie-tee-ah)

Eocene epoch (53 to 48 million years ago)
The Knightia was designated the state fossil in 1987.

During the Eocene, three massive lakes covered parts of Wyoming, Colorado, and Utah. The sedimentary remains of these lakes make up the Green River formation, and within it are the fossils of fish, birds, reptiles, plants, insects, and other flora and fauna. One of the most common fish that can be found in certain layers of the Green River formation is the Knightia. The Knightia was a fish in the same family as herring and grew up to 10 inches long. There are several fossil fish quarries near the town of Kemmerer, where visitors can pay to dig for these fish. Knightia are so common that often many specimens can be found on the same piece of rock.

Knightia specimen from a quarry in Wyoming

District of Columbia

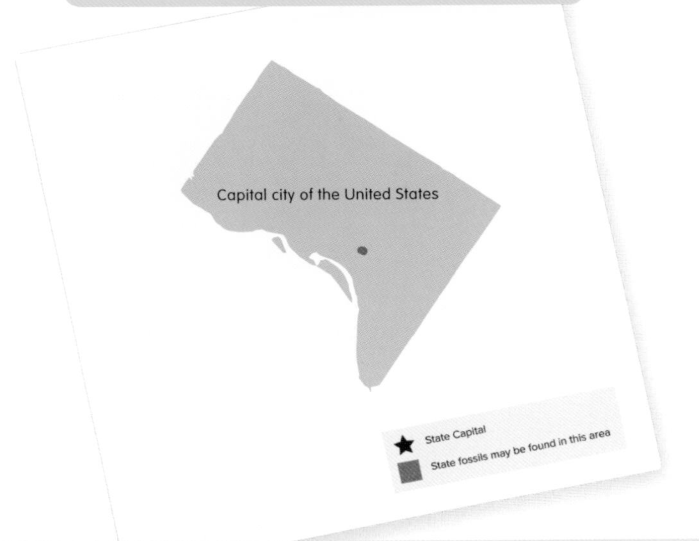

CAPITALSAURUS

Capitalsaurus
(pronounced Ca-pit-al-saw-rus)

Cretaceous period (125 to 113 million years old)
The Capitalsaurus was designated the official dinosaur of the District of Columbia in 1998.

The vertebrae of the Capitalsaurus

While the District of Columbia (Washington, DC) is a federal district and not a state, it does have an official dinosaur that deserves recognition. A dinosaur vertebra and a few bone fragments were discovered by workmen doing sewer construction on First and F Street Southeast in 1898. It was found in the Cretaceous-aged Arundel formation and was called *Creosaurus potens* (pronounced Kree-oh-saw-rus poh-tens). Later it was renamed *Dryptosaurus potens* (pronounced Drip-toh-saw-rus poh-tens) before being informally adopted as Capitalsaurus. Little is known about Capitalsaurus other than it was a theropod dinosaur. In 1998, the Capitalsaurus was designated DC's official dinosaur, and the site of its discovery is now called Capitalsaurus Court.

GLOSSARY

Absolute Dating: Finding specifically how old a rock or fossil is.

Amber: Fossilized tree sap or resin.

Arthropod: An invertebrate animal with a segmented body, an exoskeleton, and limbs with joints. Insects, crabs, and trilobites are examples.

Bivalve: A marine mollusc with two hinged shells. Mussels, clams, and oysters are examples of bivalves.

Calyx: Central part of the crinoid containing digestive and reproductive organs.

Canines (teeth): Large sharp teeth or fangs.

Carbon: A chemical element that is an important component of life.

Carnivore: An animal that only eats other animals.

Cast: An object created by filling a mold with a substance that hardens.

Cephalopod: An animal from the class Cephalopoda. They are invertebrates and usually have many arms. Squids, octopuses, and belemnites are examples.

Clade: A group of organisms with a common ancestor.

Coprolites: Fossilized poop, dung, or droppings.

Dinosaurs: Group of prehistoric reptiles that lived on land during the Mesozoic era and share a common leg structure (erect hind limbs). Tyrannosaurus rex, Triceratops, Stegosaurus, Allosaurus, and Hadrosaurus are all examples of dinosaurs.

Echinoderm: Animal of the phylum Echinodermata. They are marine invertebrates and have 5-point symmetry. Sea stars, crinoids, and sea urchins are examples.

Endocast: The cast made from the inside of a hollow object. A dinosaur skull can be filled with plaster to find the shape of the inside; the resulting plaster object is an endocast.

Exoskeleton: The external skeleton that protects an animal. Insects, clams, and even snails have exoskeletons.

Extinction: When an entire species dies out.

Genus: A group of distinct organisms that ranks higher than species in taxonomic ranks, but below family.

Hadrosaurs: Dinosaurs of the Hadrosauridae family, also known as duck-billed dinosaurs.

Herbivore: An animal that only eats plants and vegetation.

Ice Age: A period of time when the Earth's temperature was lower than normal, which resulted in large glaciers covering a lot of the Earth's surface. The most famous Ice Ages were during the Pleistocene.

Ichnotaxon: Name for a group of trace fossils.

Incisors: Teeth designed for cutting.

Index fossil: A fossil that existed for a short period of time and so is useful for determining the age of the rock it is in.

Invertebrates: Animals that do not have a backbone.

Mammal: An animal of the Mammalia class. Mammals are warm blooded vertebrates, have hair, and females produce milk to feed their young. Humans, Mammoths, and horses are examples of mammals.

Mold: Impression of a fossil made after the original fossil deteriorates.

Molluscs: An animal of the *Mollusca phylum*. Molluscs are invertebrates and some examples are squids, slugs, and snails.

Organism: A lifeform. All living things are organisms.

Paleontologist: A scientist who studies fossils and prehistoric life.

Paleontology: The study of fossils and prehistoric life.

Permafrost: A layer of earth and rock that is frozen for more than two years, sometimes tens of thousands of years.

Petrified Wood: Fossilized wood, usually preserved by permineralization so that it has some of its original form.

Phylum: A group of distinct organisms that ranks higher than class in taxonomic ranks, but below kingdom.

Polyps: The tiny invertebrate animals that create coral reefs.

Relative Dating: Finding the age of a rock or fossil in relation to the rocks above and below it.

Sauropod: Group of long-necked herbivore dinosaurs. Diplodocus, Apatosaurus, and Sauroposeidon are examples.

Sediment: Matter that settles to the bottom of a liquid. Sand in a river is an example of sediment.

Spores: A cell that can grow into a new organism.

Suspension feeder: An organism that feeds on tiny organic material in water.

Tar: A sticky dark liquid form of petroleum.

Telson: The last exoskeleton segment on some arthropods. Horseshoe crabs and Eurypterids have telsons.

Theropods: Group of mostly carnivorous dinosaurs that walked on two legs. The *Tyrannosaurus rex* and Allosaurus are theropods. Birds are believed to be descended from theropods.

Vertebra: A back bone.

Vertebrates: Animals with back bones.

PLACES TO SEE FOSSILS

Fossils can be found in every state, and the best places to see them are in museums and parks. While there are many museums and parks in every state, the ones listed here have fossils on display or are generally related to fossils. There are many more museums and parks not listed, so ask around! Be sure to check visiting hours and information before you go.

ALABAMA

Alabama Museum of Natural History - Tuscaloosa
Anniston Museum of Natural History - Anniston
Mann Wildlife Learning Museum - Montgomery
McWane Science Center - Birmingham

ALASKA

Alaska Museum of Science & Nature - Anchorage
Alaska State Museum - Juneau
Pratt Museum - Homer
University of Alaska Museum of the North - Fairbanks

ARIZONA

Arizona Museum of Natural History - Mesa
Arizona-Sonora Desert Museum - Tucson
Museum of Northern Arizona - Flagstaff
Petrified Forest National Park

ARKANSAS

Arkansas Museum of Natural Resources - Smackover
Arkansas State University Museum - Jonesboro
Museum of Discovery - Littlerock
Turner Neal Museum of Natural History and Pomeroy Planetarium - Monticello

CALIFORNIA

Buena Vista Museum of Natural History & Science - Bakersfield
California Academy of Science - San Francisco
Children's Natural History Museum - Fremont
Fallbrook Gem & Mineral Museum - Fallbrook
Gateway Science Museum - California State University, Chico
Humboldt State University Natural History Museum - Arcata
La Brea Tar Pits & Museum - Los Angeles
Maturango Museum - Ridgecrest
Morro Bay State Park Museum of Natural History - Morro Bay
Natural History Museum Los Angeles County - Los Angeles
Pacific Grove Museum of Natural History - Pacific Grove
Petaluma Wildlife & Natural Science Museum - Petaluma
Ralph B. Clark Regional Park - Buena Park
San Diego Natural History Museum - San Diego
Santa Barbara Museum of Natural History - Santa Barbara
Sierra College Natural History Museum - Rocklin
Western Science Center - Hemet
World Museum of Natural History - La Sierra University - Riverside

COLORADO

Colorado School of Mines Geology Museum - Golden
Denver Museum of Nature & Science - Denver
Dinosaur Ridge - Morrison
Florissant Fossil Beds National Monument - Florissant
Fort Collins Museum of Discovery - Fort Collins
Morrison Natural History Museum - Morrison
Museum of Western Colorado's Dinosaur Journey Museum - Fruita
Rocky Mountain Dinosaur Resource Center - Woodland Park
University of Colorado Museum of Natural History - Boulder
University of Colorado South Denver - Lone Tree

CONNECTICUT

Bruce Museum - Greenwich
Dinosaur State Park - Rocky Hill
Yale Peabody Museum of Natural History - New Haven

DELAWARE

Delaware Museum of Natural History - Wilmington

DISTRICT OF COLUMBIA

Smithsonian National Museum of Natural History

FLORIDA

Bailey-Matthews National Shell Museum - Sanibel
Florida Museum of Natural History - Gainesville
Museum of Arts & Sciences - Daytona Beach
South Florida Museum - Bradenton

GEORGIA

Fernbank Museum of Natural History - Atlanta
Georgia Southern Museum - Statesboro
Tellus Science Museum - Cartersville

HAWAII

Bernice Pauahi Bishop Museum - Honolulu

IDAHO

Hagerman Fossil Beds National Monument - Hagerman
Idaho Museum of Natural History - Pocatello
Orma J. Smith Museum of Natural History - Caldwell

ILLINOIS

Burpee Museum of Natural History - Rockford
Elgin Public Museum of Natural History & Anthropology - Elgin
Field Museum of Natural History - Chicago
Illinois State Museum - Springfield
Midwest Museum of Natural History - Sycamore
Peggy Notebaert Nature Museum - Chicago

INDIANA

Falls of the Ohio State Park - Clarksville
Children's Museum of Indianapolis - Indianapolis
Indiana State Museum - Indianapolis
Joseph Moore Museum - Richmond

IOWA

Fossil & Prairie Park Preserve - Rockford
Putnam Museum and Science Center - Davenport
University of Iowa Museum of Natural History - Iowa City

KANSAS

Fick Fossil & History Museum - Oakley
Johnston Geology Museum - Emporia
Keystone Gallery - Scott City
KU Biodiversity Institute and Natural History Museum - Lawrence
Museum at Prairiefire - Overland Park
Museum of World Treasures - Wichita
Sternberg Museum of Natural History - Hays

KENTUCKY

Ben E. Clement Mineral Museum - Marion
Big Bone Lick State Historic Site - Union
Kentucky Science Center - Louisville

LOUISIANA

Lafayette Science Museum - Lafayette
Louisiana Art & Science Museum - Baton Rouge

MAINE

Maine State Museum - Augusta
Nylander Museum - Caribou

MARYLAND

Calvert Cliffs State Park - Lusby
Calvert Marine Museum – Solomons
Dinosaur Park – Laurel

MASSACHUSETTS

Beneski Museum of Natural History - Amherst
Berkshire Museum - Pittsfield
Ecotarium - Worcester
Harvard Museum of Natural History - Cambridge

MICHIGAN

Central Michigan University Museum of Cultural and Natural History - Mount Pleasant
Cranbrook Institute of Science - Bloomfield Hills
Kingman Museum - Battle Creek

Michigan State University Museum - East Lansing
University of Michigan Museum of Natural History - Ann Arbor

MINNESOTA

Bell Museum of Natural History - Minneapolis
Science Museum of Minnesota - Saint Paul

MISSISSIPPI

Dunn-Seiler Museum - Starkville
Mississippi Museum of Natural Science - Jackson

MISSOURI

Branson Dinosaur Museum - Branson
Joplin History & Mineral Museum - Joplin
St. Louis Science Center - St. Louis

MONTANA

Carter County Museum - Ekalaka
Garfield County Museum - Jordan
Great Plains Dinosaur Museum - Malta
Museum of the Rockies - Bozeman
Two Medicine Dinosaur Center - Bynum

NEBRASKA

Agate Fossil Beds National Monument - Harrison
Hastings Museum - Hastings
Trailside Museum of Natural History - Crawford
University of Nebraska State Museum - Lincoln

NEVADA

Berlin-Ichthyosaur State Historic Park - Austin
Nevada State Museum - Carson City
Nevada State Museum - Las Vegas
Las Vegas Natural History Museum - Las Vegas
Museum of Natural History, University of Nevada - Reno
Tule Springs Fossil Beds National Monument - Las Vegas

NEW HAMPSHIRE

Children's Museum of New Hampshire - Dover
The Little Nature Museum - Warner

NEW JERSEY

Morris Museum - Morristown
Newark Museum - Newark
New Jersey State Museum - Trenton
Rutgers University Geology Museum - New Brunswick

NEW MEXICO

Las Cruces Museum of Nature & Science - Las Cruces
Mesalands Community College's Dinosaur Museum - Tucumcari
New Mexico Museum of Natural History & Science - Albuquerque
Prehistoric Trackways National Monument - Las Cruces
Ruth Hall Museum of Paleontology - Abiquiu

NEW YORK

American Museum of Natural History - New York City
Buffalo Museum of Science - Buffalo
Museum of the Earth - Ithaca
New York State Museum - Albany
Penn Dixie Fossil Park & Nature Preserve - Blasdell

NORTH CAROLINA

Asheville Museum of Science - Asheville
Aurora Fossil Museum - Aurora
Cape Fear Museum - Wilmington
Museum of Life + Science - Durham
North Carolina Museum of Natural Sciences - Raleigh
Schiele Museum of Natural History - Gastonia

NORTH DAKOTA

Dickinson Museum Center - Dickinson
North Dakota Heritage Center & State Museum - Bismarck

OHIO

Cincinnati Museum of Natural History & Science - Cincinnati
Cleveland Museum of Natural History - Cleveland
Fossil Park - Sylvania
Limper Geology Museum - Oxford

OKLAHOMA

Sam Noble Museum of Natural History - Norman

OREGON

John Day Fossil Beds National Monument - Kimberley
Museum of Natural and Cultural History - Eugene
Oregon Paleo Lands Institute - Fossil
Rice Northwest Museum of Rocks and Minerals – Hillsboro

PENNSYLVANIA

The Academy of Natural Sciences of Drexel University - Philadelphia
Carnegie Museum of Natural History - Pittsburgh
Delaware County Institute of Science - Media
Earth and Mineral Sciences Museum & Art Gallery - University Park
Everhart Museum - Scranton
State Museum of Pennsylvania - Harrisburg
The Wagner Free Institute of Science - Philadelphia

RHODE ISLAND

Roger Williams Park Museum of Natural History - Providence

SOUTH CAROLINA

The Bob Campbell Geology Museum - Clemson
The Charleston Museum - Charleston
McKissick Museum - Columbia
South Carolina State Museum - Columbia

SOUTH DAKOTA

Badlands National Park
The Journey Museum & Learning Center - Rapid City
South Dakota School of Mines & Technology Geology Museum - Rapid City

TENNESSEE

Discovery Park of America – Union City
Gray Fossil Site & Museum - Gray
McClung Museum of Natural History & Culture – Knoxville

TEXAS

Brazos Valley Museum of Natural History - Bryan
Brazosport Museum of Natural Science - Chute
Fort Worth Museum of Science and History - Fort Worth
Heard Natural Science Museum & Wildlife Sanctuary - McKinney
Houston Museum of Natural Science - Houston
Mayborn Museum - Waco
Mineral Wells Fossil Park - Mineral Wells
Perot Museum of Nature and Science - Dallas
Texas Memorial Museum - Austin
Waco Mammoth National Monument - Waco

UTAH

Box Elder Museum of Natural History - Brigham City
BYU Museum of Paleontology - Provo
The Dinosaur Museum - Blanding
Dinosaur National Monument - Jensen
John Hutchings Museum of Natural History - Lehi
Museum of Ancient Life at Thanksgiving Point - Lehi
Museum of Moab - Moab
Museum of the San Rafael Swell - Castle Dale
Natural History Museum of Utah - Salt Lake City
St. George Dinosaur Discovery Site - St. George
Utah Field House of Natural History State Park Museum - Vernal
Utah State University Eastern Prehistoric Museum - Price

VERMONT

Montshire Museum of Science - Norwich
Mount Holly Historical Community Museum - Belmont
The Perkins Geology Museum at the University of Vermont - Burlington

VIRGINIA

Virginia Museum of Natural History - Martinsville
Virginia Tech Museum of Geosciences - Blacksburg

WASHINGTON

Burke Museum of Natural History and Culture - Seattle
Imagine Children's Museum - Everett
Stonerose Interpretive Center & Eocene Fossil Site - Republic

WEST VIRGINIA

West Virginia Geological & Economic Survey Museum - Morgantown

WISCONSIN

Dinosaur Discovery Museum - Kenosha
Kenosha Public Museum - Kenosha
University of Wisconsin Geology Museum - Madison
University of Wisconsin-Stevens Point Museum of Natural History - Stevens Point

WYOMING

Fossil Butte National Monument - Kemmerer
Tate Geological Museum - Casper
University of Wyoming Geological Museum - Laramie
Wyoming Dinosaur Center – Thermopolis

FURTHER READING

To learn more about state fossils and paleontology in general, check out the following books and websites:

Brusatte, Stephen. *Stately Fossils: A Comprehensive Look at the State Fossils and Other Official Fossils*. Boulder, CO: Fossil News, 2002.

Case, Gerald. *A Pictorial Guide to Fossils*. Malabar, FL: Krieger Publishing Company, 1992.

Fortey, Richard. *Fossils: The Key to the Past*. Cambridge, MA: Harvard University Press, 1991.

McPherson, Alan. *State Geosymbols: Geological Symbols of the 50 United States*. Bloomington, IN: AuthorHouse, 2011.

Paul, Gergory. *The Princeton Field Guide to Dinosaurs*. Princeton, NJ: Princeton University Press, 2010.

Walker, Cyril A., and David Ward. *Fossils*. New York: DK, 2002.

WEBSITES:

http://www.ucmp.berkeley.edu/
-University of California Museum of Paleontology

http://www.thefossilforum.com/
-The Fossil Forum

http://www.myfossil.org/
-MyFOSSIL | Social Paleontology

http://www.fossilguy.com/
-Celebrating the Richness of Paleontology through Fossil Hunting

http://gatorgirlrocks.com/
-Gator Girl Rocks: America's Best Rockhounding Resource

http://oceansofkansas.com/
-Oceans of Kansas

https://mostlymammoths.wordpress.com/
-Mostly Mammoths, Mummies, and Museums

https://en.wikipedia.org/wiki/Fossil
-Wikipedia entry about fossils

http://www.ereferencedesk.com
-50 States Guide

PHOTO CREDITS

Cast and mold: Photo by author
Permineralization: Jake Harper Photo 2017. Specimen courtesy of Ed Rigel, collector & preparer
Mineral replacement: Photo by author
Carbonizaton: Photo by author
Trace fossil: Photo by author. Specimen courtesy of Dinosaur State Park, Rocky Hill, CT
Original remains: Photo by author
Fossil hunting: Photo by author
Alabama: Courtesy of John Friel, Alabama Museum of Natural History
Alaska: Courtesy of GeoDecor Inc.
Arizona: Jake Harper Photo 2017. Specimen courtesy of Ed Rigel, collector & preparer
Arkansas: Courtesy of the University of Arkansas Museum Collections
California: Photo by author. Specimen courtesy of the Department of Geosciences, Princeton University
Colorado: Courtesy of Brock Sisson, Fossilogic LLC
Connecticut: Photo by author. Specimen courtesy of Dinosaur State Park, Rocky Hill, CT
Delaware: Photo by author. Specimen courtesy of Gene Hartstein
Florida: Photo by Heather Spence. Specimen courtesy of Karl E. Limper Geology Museum, Miami University, Ohio. Donated by Richard Polsky
Georgia: Photo by author
Hawaii: Courtesy of Molly Hagemann, Bernice Pauahi Bishop Museum
Idaho: Photo by Kathy Neenan. Specimen courtesy of the Arizona Museum of Natural History
Illinois: Courtesy of the Rob Coleman collection
Indiana: Photo by author
Iowa: Photo by author
Kansas: Courtesy of KU Natural History Museum
Kentucky: Photo by author
Louisiana: Photo by author
Maine: Courtesty of R.A. Gastaldo, Whipple-Coddington Professor of Geology, Colby College
Maryland: Photo by author
Massachusetts: Photo by author. Specimen courtesy of the Beneski Museum of Natural History, Amherst College
Michigan: Photo by author. Specimen courtesy of the Beneski Museum of Natural History, Amherst College
Minnesota: Photo by author
Mississippi: Photo by author. Specimen photographed at George Mason University
Missouri: Courtesy of Guy Darrough
Montana: Courtesy of Korite International Inc
Nebraska: Photo by author. Specimen courtesy of the Beneski Museum of Natural History, Amherst College
Nevada: Photo by Tom Dyer. Specimen courtesy of Nevada State Museum, Las Vegas
New Hampshire: Courtesy of Gary D. Johnson
New Jersey: Photo by author. Specimen courtesy of the New Jersey State Museum
New Mexico: ©Bailey Library and Archives, Denver Museum of Nature & Science
New York: Courtesy of the Rob Coleman collection
North Carolina: Photo by author
North Dakota: Photo courtesy of anonymous
Ohio: Photo by Heather Spence. Specimen courtesy of Karl E. Limper Geology Museum. Miami University, Ohio. From the collection of Walter Gross III
Oklahoma: Drawing by Brock Sisson, Fossilogic LLC
Oregon: Courtesy of the Stonerose Interpretative Center
Pennsylvania: Courtesy of Kerry Matt
Rhode Island: Courtesy Museum of Natural History and Planetarium, Roger Williams Park, Providence, Rhode Island
South Carolina: Photo by author. Specimen courtesy of the Beneski Museum of Natural History, Amherst College
South Dakota: Courtesy of Dinosaur Resource Center, Woodland Park, Colorado
Tennessee: Courtesy of James Davison
Texas: Drawing by Brock Sisson, Fossilogic LLC
Utah: Specimen courtesy of the Department of Geosciences, Princeton University
Virginia: Photo by author.
Vermont: Whale photo by Bill Dillilo UVM Photo. Specimen courtesy of UVM Geology Department. Mammoth tusk photo by Robin Eatmon. Specimen courtesy of the Mount Holly Community Historical Museum.
Washington: Photo by author. Specimen courtesy of the Beneski Museum of Natural History, Amherst College
West Virginia: Photo by author. Specimen courtesy of New Jersey State Museum
Wisconsin: Photo by author. Specimen courtesy of the collection of John and Dorothy Stade
Wyoming: Photo by author. Specimen courtesy of Lindgren Fossils
District of Columbia: Modified from Gilmore, Charles W., *Osteology of the Carnivorous Dinosauria in the United States National Museum: With Special Reference to the Genera Antrodemus (Allosaurus) and Ceratosaurus.* Washington, DC: Government Printing Office, 1920. https://archive.org/details/osteologyofcarni00gilm

ACKNOWLEDGMENTS

This book would not have been possible without the recommendations, contacts, information, editing, and good will of the following people:

Tracie Bennitt
Eric Biddle
Robert Coleman
Guy Darrough
James Davison
Robert Detrich
Guy DiTorrice
Megan Duffy-Johnson
Janet Dunkelberger
Robin Eatmon
Meg Enkler
John Friel
Henry Galiano
Ray Garton
Robert Gastaldo
Sam Ohu Gon III
Molly Hagemann
Eugene Hartstein
Kendall Hauer
Jennifer Humphrey
John Issa
Russell Jacobson
Gary Johnson
Andreas Kerner
Michael Kieron
Mike Levy
Thomas Lindgren
Tony Lindgren
Adelina Longoria
Kerry Matt
Katherine Meade
Kathy Neenan
Christian Nelligar
Johanna Nelligar
Cindy North and Jake
Rene O'Connell
Lauren Orsini
David Parris
David Penney
Sheryl Robas
Ed Rigel
Elyse Rodriguez
Owen Rodriguez
Donna Russell
Imelda Sarnowiec
Brock Sisson
Joshua Smith
Thom Smith
Heather Spence
John and Dorothy Stade
Elizabeth Stonick
Mary Suter
Jeanne Timmons
Michael Triebold
Mark Uhen
Sali Underwood
Alfred Venne
Matthew Vuolo
Sarah Wang
Hilary-Morgan Watt
Cheryl Weber
Andrew Wendruff
George Winters
Brandy Zzyzx
Justin Zzyzx

INDEX